WATER SOFTENING WITH POTASSIUM CHLORIDE

WATER SOFTENING WITH POTASSIUM CHLORIDE
Process, Health, and Environmental Benefits

William Wist
Jay H. Lehr
Rod McEachern

A JOHN WILEY & SONS, INC., PUBLICATION

Library of Congress Cataloging-in-Publication Data:

Wist, William (1945–2007)
    Water softening with potassium chloride : process, health, and environmental benefits /
William Wist, Jay H. Lehr, Rod McEachern.
        p. cm.
    Includes index.
    ISBN 978-0-470-08713-8 (cloth)
    1. Water–Softening. 2. Potassium chloride–Environmental aspects. 3. Water–Purification.
4. Ion exchange. I. McEachern, Rod, 1958– II. Lehr, Jay H., 1936– III. Title.
    TD466.W57 2009
    628.1′666–dc22

                                                                    2009013333

10 9 8 7 6 5 4 3 2 1

This book is
for the devotion a

WATER SOFTENING WITH POTASSIUM CHLORIDE
Process, Health, and Environmental Benefits

William Wist
Jay H. Lehr
Rod McEachern

A JOHN WILEY & SONS, INC., PUBLICATION

Published by John Wiley & Sons, Inc., Hoboken, New Jersey
Published simultaneously in Canada

For general information on our other products and services or for technical support, please contact our Customer Care Department within the United States at (800) 762-2974, outside the United States at (317) 572-3993 or fax (317) 572-4002.

Wiley also publishes its books in a variety of electronic formats. Some content that appears in print may not be available in electronic formats. For more information about Wiley products, visit our web site at www.wiley.com.

Library of Congress Cataloging-in-Publication Data:

Wist, William (1945–2007)
    Water softening with potassium chloride : process, health, and environmental benefits / William Wist, Jay H. Lehr, Rod McEachern.
        p. cm.
    Includes index.
    ISBN 978-0-470-08713-8 (cloth)
    1. Water–Softening.   2. Potassium chloride–Environmental aspects.   3. Water–Purification.
4. Ion exchange.   I. McEachern, Rod, 1958–   II. Lehr, Jay H., 1936–   III. Title.
    TD466.W57 2009
    628.1′666–dc22

                                                                          2009013333

10 9 8 7 6 5 4 3 2 1

This book is dedicated to the memory of Bill Wist,
for the devotion and love he gave to his family, his research,
and his many friends.

CONTENTS

PREFACE

Practical problems associated with hard water, including scale buildup and the precipitation of soap films, have plagued humanity for thousands of years. The introduction of ion exchange resins in the early 1900s offered the first practical means of reducing water hardness, and water softening with cation exchange resins quickly became widespread. The introduction in 1944 of ion exchange resins based on the copolymerization of polystyrene and divinylbenzene (DVB) brought water softening technology to a state of maturity, and use of this resin became the standard method of water softening worldwide.

One common feature of water softening by ion exchange was the use of common salt (sodium chloride, NaCl) as regenerant. For many years, sodium chloride was the preferred material for this purpose because of its effectiveness and low cost. However, as the use of home and industrial water softeners spread, the large quantities of spent regenerant released to the environment became a concern. Research was therefore initiated to evaluate the potential use of potash (potassium chloride, KCl) as an alternative to sodium chloride for use as a regenerant in softening applications. Potassium chloride is widely used as a fertilizer, so its use in softening offered the potential of a "win-win" situation in which the release of sodium to the environment was reduced, while simultaneously providing an effluent that could be used as a fertilizer.

At PotashCorp, a research project to evaluate application of potassium chloride as a regenerant was initiated in the late 1980s. Research performed by Bill Wist showed the effectiveness of KCl as a regenerant and compared its performance to NaCl in critical areas such as solubility, capacity, solubility speed and cycle times. The high quality research patiently done by Bill over many years led to the successful commercialization of KCl as a water softening regenerant under the trade name "SoftTouch."

Throughout the research and development phase of the SoftTouch project, Bill Wist was the project leader. Bill provided the technical leadership, developed the test protocols, performed the tests, and effectively communicated the results to PotashCorp sales staff, water quality specialists, and the public. As testament to his dedication, Bill even installed a softener with KCl regenerant in his own home, and performed household water softening experiments as part of the test program. As the use of KCl in water softening spread, Bill answered many technical questions for water softening dealers and service people; he also communicated directly with customers. Bill's unique expertise in this field, his practical problem solving abilities, along with his easy-going personality earned him the reputation as the "go-to" person in the field of KCl for water softening.

In his retirement, Bill Wist initiated the idea for this book, as a means of educating the public about the technology of water softening and the effectiveness and environmental benefits of the use of KCl as a regenerant. In addition to his work as researcher, Bill also organized the entire text and wrote several of the chapters.

We hope that you find this text interesting and useful as a reference for water softening applications. This text, along with the widespread use of KCl as a regenerant, stands as a legacy to the patient hard work and innovative research performed by Bill Wist.

ACKNOWLEDGMENTS

The research and development described in this text was performed while I was employed at the PotashCorp Technical Services pilot plant. I would like to thank PotashCorp management, who supported the SoftTouch project through the many years of research, development, and customer support. Specifically, the interest, support, and technical leadership of Rick Lacroix (VP of Technical Services), Graeme Strathdee (Director of R&D), and Al Mulhall (Manager of R&D) is appreciated.

One of the more gratifying results of the KCl research was to witness the emerging use of spent regenerant solutions as graywater for irrigation. The KCl content of this water provides nutrition for plants, which is superior to the past practice of disposing of the NaCl waste to the environment. The contributions of Kim Polizotto to understanding the agronomy and soil science associated with the use of KCl-rich graywater is appreciated. Kim, along with Sid Blair, should also be recognized as key members of the team that helped to spread the word on the use of KCl regenerant.

Thanks are also extended to Jay Lehr and Rod McEachern for their contributions to the completion of this book.

The excellent technical work on the illustrations done by Patrick Breton, and the support of Creative Fire are gratefully appreciated.

The many years of research required thousands of chemical analyses, all of which were performed by Jana Nguyen. The hard work, dedication, and always-positive attitude of Jana is gratefully acknowledged.

Finally, I am grateful to my wife Donna, and our children, Darin, Kelly, and Jolene, along with our grandchildren and other family members, whose love and support meant so much to me during the many years of research described in this text.

BILL WIST
June 25, 2007

CHAPTER 1

WHAT IS POTASSIUM CHLORIDE?

Potassium is the seventh most abundant element and makes up about 2.6% of the Earth's crust. Potassium metal is far too reactive to be found uncombined in nature, so it exists as compounds in minerals. Examples of compounds are potassium silicate, potassium nitrate, potassium sulfate, potassium hydroxide, and, of course, potassium chloride. Potassium silicate, or feldspar [$KAlSi_3O_8$], is the most abundant compound of potassium on Earth, but feldspar minerals are very stable, so extraction of K^+ from feldspars is not industrially viable. Potassium chloride (KCl), which is also abundant in nature, is better suited to separation and concentration of potassium on an industrial scale. In industry, KCl is used as a feedstock for the production of many other potassium compounds. Potassium chloride is also the most important compound for supplying potassium to the fertilizer industry.

Potassium deposits are spread throughout the world in minerals such as sylvite (KCl), sylvinite (KCl/NaCl), langbeinite [$K_2Mg_2 (SO_4)_3$], and carnallite ($KCl \cdot MgCl_2 \cdot 6H_2O$). These deposits are found in ancient, shallow saline lake and sea beds.

The world's largest reserve of potassium is located in west-central Canada, primarily in the province of Saskatchewan. The potash deposits formed over 380 million years ago as a result of the final stage of evaporative concentration of seawater in the Middle Devonian Sea. This shallow, inland sea extended from the southern part of the Northwest Territories southeast through Alberta, through southern Saskatchewan into Manitoba and southeast into North Dakota. Potash salts are concentrated in the southeastern extent of this basin, underlying much of southern Saskatchewan. The average depth of these deposits is ~1 kilometer or 3300 feet. The mineral name for the predominant potash ore is sylvinite, which is a mixture of halite ~60% (sodium chloride or NaCl), sylvite ~35% (potassium chloride or KCl), and clay (and other water-insoluble minerals) ~5%. The concentration of the minerals in sylvinite can vary significantly depending on the location of the potash ore.

SASKATCHEWAN POTASH HISTORY

The discovery of the potash deposits in Saskatchewan occurred in 1943 from an exploratory oil well in southern Saskatchewan. The first core sample was at a depth of 2 kilometers or 6600 feet. Another core sample from an exploratory oil well northwest of the first core in west central Saskatchewan discovered a high-grade potassium ore at a depth of 1 kilometer. Exploration was expanded over the next few years, and in the 1950s it was determined that a vast (probably the world's largest) potash deposit was below the surface of most of southern Saskatchewan.

By conservative estimates, potash mineralization in Saskatchewan could supply the world potassium demand for several hundred years.

The first potash mine in Saskatchewan was completed in 1958. However, the shaft was flooded the same year by a high-pressure sand- and water-bearing formation above the potash zone, and the mine did not return to production until 1965. The ~100-meter-thick layer of water-bearing sands of the Lower Cretaceous Mannville Group occurs about halfway between the surface and the potash zone; within the Saskatchewan potash industry the common name for this unit is the Blairmore Formation. The water pressure in this formation is proportional to burial depth, so it can be as high as 500 psi. To successfully sink a shaft to the potash zone, Blairmore sands had to be sealed. A successful technique was developed in the 1960s in which the Blairmore formation was frozen until the shaft was dug and sealed with concrete and iron-ring tubbing segments to form a permanent watertight shaft lining. The second potash mine began production of potash mining 1000 meters below the surface in 1962. By 1974 there were 10 potash

TABLE 1-1: Production capacity (in millions of metric tons per year, as KCl) of Saskatchewan potash mines, as of 2006

Producer	Location	Mining Method	Nameplate Capacity (MTPY)
Agrium	Vanscoy	Conventional	1.750
Mosaic	Belle Plaine	Solution	2.533
Mosaic	Colonsay	Conventional	1.485
Mosaic	Esterhazy K1/K2	Conventional	3.928
PotashCorp	Allan	Conventional	1.885
PotashCorp	Cory	Conventional	1.361
PotashCorp	Lanigan	Conventional	3.828
PotashCorp	Patience Lake	Solution	1.033
PotashCorp	Rocanville	Conventional	3.044

mines operating in Saskatchewan, with a total annual production capacity of 13 million tons of potassium chloride. Nine of the mines used conventional mining techniques, and one used a solution mining technique that does not require the sinking of a shaft.

The original 10 mines built in Saskatchewan in the 1960s were privately owned by nine separate companies. At present, all 10 mines are still operating, but there are only three separate companies. The original mine built at Patience Lake in 1958 was flooded in 1988 and has since been converted to a solution mine. Many expansions have taken place over the years, so that annual production capacity currently exceeds 20 million tons.

The Saskatchewan potash industry is currently expanding rapidly in an effort to meet rising world demand. Production from the Saskatchewan potash mines exceeds 20 million tons annually (Table 1-1) and provides over 30% of the worlds potash. Data shown in Table 1-1 apply to Saskatchewan mines only; some of the producers have potash capacity elsewhere in the world, but these other facilities are excluded from this table.

POTASH MINING

The deposits in Saskatchewan are typical of many potash deposits around the world, that is, they were formed millions of years ago by evaporation of an inland sea. When an inland sea is isolated from the ocean, the sea salts in the water will become more concentrated (saturated) because of evaporation. If the climate is warm and dry, then the sea salts will over time become saturated, so that further evaporation results in precipitation (crystallization) of the salts. Sodium chloride is the least soluble of the common salts in ocean water, so it crystallizes out from the solution first. Other salts, including potash (KCl) crystallize out later as they too achieve saturation. As the various sea salts crystallize out and fall to the bottom of the inland sea, a layer of these salts is built. The various salts crystallize in sequence (according to their quantity and solubility), so there are discrete layers of KCl-rich salts that can be mined economically. Potash deposits, formed by crystallization of ancient seas, can be covered by sediment and other material over geological time, so that the potash deposits in modern times can be buried at substantial depths. The Saskatchewan deposits vary in depth below surface. Most of the mines in the province are conventional mines, located near the northern (shallow) end of the potash deposit; for these mines the potash is located approximately 1 kilometer (3300 feet) below the surface. Once the potash ore is mined and hoisted to the surface, the KCl is then separated from the other salts and sold, primarily as a plant nutrient (fertilizer). Potash production is therefore the recovery and purification of material from sea salts!

The potash deposits in Saskatchewan were formed at the bottom of an ancient inland sea 380 million years ago, and therefore the salts were deposited over a large area, in a relatively flat, thick layer of salt-rich material. Many such deposits were formed around the world in a similar way. In some cases the potash deposit became tilted or folded over time because of geological processes. In Saskatchewan, fortunately, the potash deposits remain relatively undisturbed and flat, which makes it easier to recover the potash ore safely and efficiently.

Potash ore is a mixture of KCl (sylvite) and NaCl (halite) along with minor amounts of $KCl \cdot MgCl_2 \cdot 6H_2O$ (carnallite) and water-insoluble minerals. The mixture of sylvite and halite is given the mineralogical name "sylvinite." Sylvinite, like sodium chloride, is a fairly hard, but brittle material. Therefore, mining equipment has been developed to efficiently extract the ore by taking advantage of this characteristic. Most of the potash mining is done with large continuous boring equipment, originally developed for coal mining. A typical continuous borer consists of large rotors mounted on a powerful track-driven tractor (Figs. 1-1 and 1-2). The tractor pushes the rotors against the mining face (ore body), and the rotors scrape the potash ore from the face. The particle size of the ore extracted in this manner will vary from 1-foot diameter all the way down to dust-sized particles. The potash falls to the ground, where it is scooped up by the mining machine and deposited onto a conveyor belt. The circular motion of the rotors leaves a pattern of characteristic circular grooves on the ore face (Fig. 1-3).

FIGURE 1-1 A potash miner inspects the front of a continuous boring machine. Machines such as this unit will cut up to 700 metric tons of ore per hour from the ore body and deposit it on a conveyor belt for transport to the shaft.

The ore cut by the boring machine is deposited directly onto a movable (extensible) conveyor belt, which delivers it to a permanent belt. There can be as many as 25 miles of underground conveyor belts in the larger mines, transporting up to 2000 tons per hour of ore to the shaft. Inside the shaft is equipment that loads the ore into specially designed containers ("skips"). Each skip holds 20–30 tons of ore, which is rapidly hoisted to the surface. The skips make many cycles from underground to surface each hour; the total capacity for hoisting ore is approximately 1000 to 2000 tons per hour. Once on the surface, the ore is transported to the refinery for processing, as described in the following sections.

In a typical conventional potash mine the majority of the ore is extracted with continuous boring machines as described above. Many

FIGURE 1-2 A mechanic making repairs and adjustments to the front end of a continuous boring machine in potash service.

other types of equipment are also required as part of the mining operation. Some equipment is used to install rock bolts, which provide stability to the roof of the mine and thereby reduce the risk of falling rocks. Other equipment consists of boom-mounted rotating cutting heads (Fig. 1-4), which are used to trim irregular surfaces in the mine workings. The equipment in an underground mining operation requires repair from time to time, and so each mine has underground shops with teams of welders, mechanics, and electricians (Fig. 1-5).

Many conventional underground potash mines in Saskatchewan use some variation of long room and pillar mining. The long room and pillar technique involves cutting a series of rooms, which can be 4000 to 6000 feet long and 40 to 80 feet wide. Along with conservative extraction ratios, this provides enough support to prevent the mine

FIGURE 1-3 A potash miner stands in front of the sylvinite ore face, which shows the characteristic circular grooves cut into the ore by the tungsten carbide bits on the mining machine rotors. The miner is holding a scaling bar, which is used to inspect the roof for loose rocks and safely remove them. Mounds of potash ore, recently extracted from the face, lie on the floor behind the miner.

rooms from collapsing. This mining pattern is illustrated in Figures 1-6 and 1-7. Stress-relief mining methods are used to alleviate risk in Saskatoon-area mines, where clay layers in the mine roof could render room-and-pillar methods unsafe in the absence of proper safe mining techniques. Stress-relief mining involves cutting underground openings in a pattern that creates areas of ground failure in order to deflect rock pressure away from other mine rooms, keeping these safe for extended time periods. The resulting stress-relief mining pattern is often called the herringbone pattern, for obvious reasons (see Fig. 1-8). The overall extraction ratio for either mining method ranges from 35% to 45%.

FIGURE 1-4 A mining machine with a boom-mounted cutting head. Machines such as this are use often used in potash mining to trim irregularities in the mine workings to ensure a safe working environment.

FIGURE 1-5 Underground shops where welders, mechanics, and electricians work to repair damaged equipment, so that it can be quickly returned to mining service. The shops shown in this photograph are 3200 feet (~1 kilometer) below the Earth's surface.

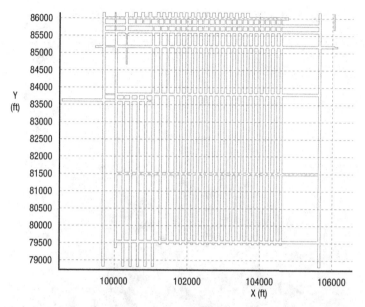

FIGURE 1-6 Example of a long room and pillar cutting pattern at the PCS Potash, Rocanville Division in southeast Saskatchewan. Coordinate axes are in mine coordinates (feet), and *X* is east and *Y* is north. Overall extraction for this panel is just under 40%.

These conservative extractions are strictly adhered to in order to minimize the likelihood of mine flooding from overlying water-bearing formations.

SOLUTION MINING

The depth of the potash ore in Saskatchewan varies. At the northern side of the deposit, the ore is found approximately 3300 feet (~1 kilometer) below the surface. However, the depth of the potash deposit is greater in the southern part of the ore body. When the ore is located significantly greater than 1 kilometer below the surface, shaft sinking and mining become more difficult and costly. Fortunately, the deeper deposits are also warmer, so it is possible to extract potash from the deeper deposits by solution mine. One such mine was built deliberately as a solution mine in the southern part of Saskatchewan, near Belle Plaine, at an ore depth greater than 1500 meters. A second mine near Saskatoon was converted to a solution mine in 1991 after it was forced to shut down because of uncontrolled flooding in 1988.

FIGURE 1-7   Photo of a 4-rotor continuous mining machine at PCS Potash, Lanigan. The long room-and-pillar mining method is utilized at Lanigan; completed 4-pass mine rooms are 42′ wide and 16′ high. An extensible conveyor belt is located behind the mining machine; the ore is transported back to the shaft by conveyor belt and then hoisted to the surface.

Solution mining for potash is based on the temperature dependence of the solubility of KCl and NaCl in water. The solubility of KCl in water or brine increases as the temperature of the solution increases, but the solubility of NaCl slightly decreases (see Fig. 1-9). This is the basis for both crystallization and solution mining.

In a solution potash mine, a number of bore holes are drilled from surface down into the potash ore zone. Hot brine saturated with NaCl but unsaturated with KCl is pumped into the bore holes and into the potash bed. Since the brine is saturated with NaCl it will only dissolve KCl. The solution eventually saturates with KCl and is pumped to the surface for processing. Solution mining uses large volumes of high-

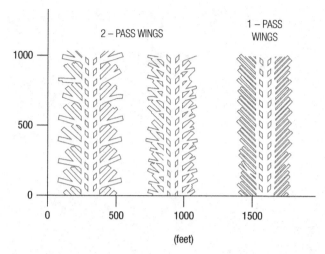

FIGURE 1-8 Examples of "herringbone" stress-relief cutting patterns employed at the PCS Potash, Allan Division in central Saskatchewan (near Saskatoon). Coordinate axes are in feet. At the present time (2008), the "1-pass wing" cutting pattern is the most common mining pattern at Allan.

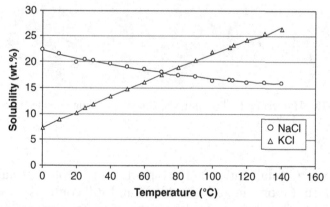

FIGURE 1-9 Solubility of KCl (—△—) and NaCl (—○—) as a function of temperature, for solutions that are saturated with respect to both salts.

temperature brine, and therefore energy consumption (and costs) can be high. The two solution mines in Saskatchewan rely on natural gas as their source of energy.

The hot brine returned to surface from a solution mine has a high content of KCl. To recover the KCl, one needs only to cool the brine and KCl crystals will precipitate (Fig. 1-9). Cooling of the brine is achieved by processing in a set of crystallizers, as described in detail in

FIGURE 1-10   Harvesting KCl from the bottom of a cooling pond using a dredge, at the PCS Patience Lake mine.

the section "Crystallization." The spent brine from crystallization remains warm (approximately 30–40 °C) and still contains a significant amount of KCl in solution. Producers therefore take advantage of the cool Saskatchewan winters, and allow the spent brine to flow through a series of specially designed open-air cooling ponds. As the brine flows through the ponds it is cooled further, and additional KCl crystallization occurs. A layer of KCl crystals grows on the bottom of the pond over time, and KCl is harvested from this layer by a dredge (Fig. 1-10) and pumped to the mill for processing. This product contains almost no clay and is white in color.

The solution mining process discussed above is used at the one mine deliberately built as a solution mine. The second solution mine in Saskatchewan is PCS Patience Lake, which was originally built as a conventional mine but was abandoned because of an uncontrolled inflow of fresh water in 1988. Production of KCl from this mine resumed in 1991 as a solution mine. The Patience Lake mine applies technology slightly different from that of a typical solution mine. In the Patience Lake process, the brine is injected warm (not hot) to reduce energy costs. The warm, KCl-rich brine returned to surface is not sent to a series of crystallizers, but instead it flows directly to a set of cooling ponds, from which KCl is recovered by dredge. The Patience Lake process allows for the recovery of KCl from a mine otherwise lost because of flooding, and it has lower energy consumption, per ton of product, than a conventional solution mine. The plant throughput from a Patience Lake type of process, however, will be lower than for a typical solution mine.

PROCESSING POTASH ORE

The ore received from underground mining consists of a mixture of KCl (30–40%), NaCl (52–69%), and water-insoluble minerals (1–8%). The basic processing requirement is removal of the NaCl and some of the insoluble material from the KCl. In the ores, the KCl and NaCl crystals are formed as discrete, single crystals of the pure compounds, rather than as a solid solution of KCl/NaCl. The individual crystals of KCl and NaCl are, however, agglomerated into a complex mechanical mixture, with clays and other insolubles located interstitially (between crystals). The one exception to this behavior is iron oxide, which is found in very low levels (parts per million) dispersed as inclusions throughout many of the KCl crystals, and imparts the characteristic rusty red colour to potash fertilizer.

Worldwide, there are many different mineral processing techniques for processing potash ore, but the most common are flotation and crystallization. Many plants rely on both of these separation techniques. For example, many plants will process the ore by flotation to produce a concentrate containing 95% KCl (suitable for sale as a fertilizer) and then process a portion of the product further, by crystallization, to produce a higher-purity product for sale as an industrial chemical. In the present discussion, we shall therefore follow the basic process for a "typical" potash refinery, treating crystallization as one of the methods of further processing the low-grade fines. It should be recognized that some plants rely solely on crystallization as their method of upgrading potash ore; in such a case the process would be different, but many of the basic concepts would be the same.

In the following sections, we examine briefly each of the key steps in the production of potash.

Ore Handling

The potash ore is hoisted to the surface in "small" (20–30 ton) containers, known as skips, as described in the section on mining. The skips will carry numerous loads from underground each hour, so that the total hoisting capacities of the plants are typically in the range of 1000 to 2000 tons per hour. The ore is dispatched from the skips onto conveyor belts (Fig. 1-11), which transport the ore to storage bins. Surface storage of ore will vary but will be in the range of 1000 to 8000 tons typically. Since potash ore is predominantly a mixture of two salts (KCl and NaCl), it is not surprising that moisture in the ore can lead to caking or setup in the storage bins (similar to what happens to the salt in a salt shaker on humid summer days). To reduce the risk of setup,

FIGURE 1-11   Transportation of potash ore from the skips (hoist) to the ore storage bin. Processing potash ore involves moving hundreds (or thousands) of tons of material from one part of the process to the next; conveyor belts are one of the common pieces of equipment used for material handling.

plant operators ensure that the contents of the storage bin are "live" or continuously moving—setup in an 8000 ton storage building is a bigger problem than in a salt shaker!

Comminution

The first task in processing potash ore is to crush the ore to the point where KCl crystals are free from NaCl and other minerals in the host rock. At this point, the KCl can be separated efficiently—in mineral processing language, one would say that the ore has been crushed to its "liberation size." The liberation size will vary for different ores. Mines in the southeast part of Saskatchewan produce an ore in which the majority of the potash crystals are liberated when crushed to less than ½ inch (12.5 mm), while in other parts of the province the ore must be crushed considerably finer.

Potash ore is moderately hard, but brittle, so it is relatively easy to crush. The first stage of crushing was traditionally done with impactors, in which the ore flows through the crusher past a series of rapidly rotating steel hammers. In recent years, there has been a tendency to replace impactors with cage mills, in which the ore is impacted with rotating steel cages rather than hammers.

The primary crushing circuit can be either open or closed. In an open circuit, the ore is screened at approximately ½-inch size, and the oversize material is fed to the crusher. The crusher discharge is then recombined with the screen undersize. In contrast, in a closed circuit, the crusher discharge is recycled back to the screen. Open circuits have the advantage of being simpler processes to operate, with less material-handling equipment. However, they lack the control on the upper particle size that a closed circuit has—a small number of particles as large as ½ inch will pass through the crushing circuit, so the downstream equipment needs to be able to process such material.

The secondary stage of comminution is generally done as a wet process. The crushed ore from the primary circuit is slurried in a brine saturated with both KCl and NaCl salts. The brine is in chemical equilibrium with the ore at the process temperatures and has a composition of approximately 10% KCl, 20% NaCl, and 70% water. The secondary stage of comminution has traditionally been done with rod mills. A rod mill is a large steel cylinder (10–15 feet in diameter and 15–20 feet long) with a large number of steel rods inside (Fig. 1-12). As the potash ore slurry flows through the rod mill, the mill turns rapidly, causing the rods to tumble. The result is a fairly gentle crush that reduces the particle size of the ore without excessive generation of fine particles (which

FIGURE 1-12 A potash plant operator standing beside a rod mill, used for the secondary comminution of potash ores.

are more difficult to process). Ore slurry from the rod mill is sent to a set of screens with ~3–4-mm openings. The ore particles with less than 3–4-mm size flow on to the next stage of the process (desliming), while the oversize (>3–4 mm) are returned to the rod mill for further size reduction.

In recent years there has been a shift away from rod mills, which are large and expensive and can be difficult to maintain. In their place, plants are installing cage mills or impactors.

Desliming

Potash ores in Saskatchewan contain 1–8% water-insoluble minerals; mines in the southeast part of the province tend toward the lower end of the range, while Saskatoon-area mines are higher. The insoluble minerals originated as suspended solids, or mud, in the ancient sea that evaporated to create the potash ore. There are a number of different minerals in the "insolubles" group, including:

— Dolomite
— Quartz
— Anhydrite
— Gypsum
— Hematite
— Illite
— Chlorite

The insoluble minerals tend to have a very small particle size of approximately 10-micron diameter. They therefore have a very large surface area that tends to adsorb the organic chemicals used later in the process, in flotation. As a consequence, it is important to do a thorough job of removing the insolubles from the potash ore, in a process known as "desliming."

The insoluble minerals are generally located interstitially in the potash ore. They are generally scrubbed free from the KCl and NaCl crystals in attrition scrubbers, in which the ore slurry is agitated vigorously for several minutes. At this point in the process, the insolubles have a very small particle size (~10 micron) while the KCl and NaCl particles are much larger (30–3000 micron). Many different methods for desliming, based on the differences in particle size, are in use, including:

— Hydrocyclones, in which the ore slurry is pumped through a specially designed conical device that collects the coarse particles from the bottom (underflow) while allowing the majority of the brine to report to the top (overflow). In a hydrocyclone the majority of the very fine materials (i.e., the insolubles) are carried along with the brine and report to the overflow, while the coarser KCl and NaCl crystals report to the underflow. A set of cyclones being used to deslime potash ore is illustrated in Figure 1-13.

— Hydroseparators, in which the fine solids (in a brine slurry) are sent to a device that allows the heavy particles (relatively large KCl and NaCl particles) to settle and be recovered from the bottom. Hydroseparators are designed so that a

FIGURE 1-13 A set of hydrocyclones, which is used for effective desliming, i.e., removal of insoluble particles, from potash ore.

substantial portion of the fluid overflows a weir at the top of the unit. The upward flow of brine carries the smallest particles (predominantly the insolubles) over the top. Thus a reasonable separation of coarse KCl/NaCl from the fine insolubles can be achieved.

— Fluidized-bed separators, in which the ore slurry is fed into a unit in which the particles are fluidized by an upward flow of clean brine. The coarse KCl/NaCl particles are recovered from the bottom of the unit, while the fine insoluble particles are recovered from the overflow.

— Slimes flotation, in which the fine solids are flocculated and then conditioned with a collector (Brogiotti and Horwald, 1975). The ore slurry then reports to a flotation cell in which the insoluble particles adhere to air bubbles and are thus transported to the top of the flotation cell, from where they are collected and removed from the ore.

The majority of the insoluble minerals have been removed from the potash ore once it has been processed in desliming. The ore is then ready for the key separation—removal of the KCl from NaCl, which is achieved in flotation.

Flotation

Froth flotation is a widely used technique for separating a wide range of valuable minerals (copper, nickel, gold, zinc, etc.) from their respective ores. The technique is based on the chemical properties of the mineral surfaces. If a mixture of minerals is slurried in a tank and air bubbles are injected, the air bubbles will adhere to those minerals that are hydrophobic (nonpolar, air-wetted), but not to the minerals that are hydrophilic (polar, water-wetted). The air bubbles will drag the desired mineral to the top of the tank, and it can be collected, separate from the undesired minerals. The process engineer can thus achieve good separation of the desired mineral by selecting appropriate conditions and surface treatment so that only the desired mineral is air-wetted.

In many industries, the surface of the desired mineral is naturally air-wetted, that is, hydrophobic. In such cases the flotation process will work with no special surface treatment. Some examples of such naturally hydrophobic materials include bitumen, graphite, sulfur, and talc. In many cases, however, like potash processing, careful surface treatment is required so that the surface of the desired mineral becomes hydrophobic (air-wetted) while the other minerals remain hydrophilic (water-wetted). A compound added to a flotation system to render the surface of the desired mineral hydrophobic is known as a *collector*. In potash processing the most common collector is a long-chain amine molecule.

The long-chain amine used in potash processing, like all common collectors, is a surfactant, that is, it has both polar and nonpolar components to the molecule. The collectors work because the polar portion of the molecule attaches to the surface of the desired mineral (KCl). When uniform (monolayer) coverage of the surface has been achieved, the nonpolar part of the collector is on the outside surface, exposed to the solution (Fig. 1-14). Once the KCl surface is coated with a monolayer of the collector, the surface has been effectively rendered hydrophobic, and air bubbles will adhere to the surface. The amine will not adhere to NaCl because of some subtle differences in the surface chemistry of these two salts; as a result, the NaCl surfaces remain hydrophilic and will not be collected by flotation.

In potash processing, the deslimed ore slurry will be conditioned with a very small quantity (e.g., 50 grams of amine per ton of ore) of the collector by mixing in a tank, launder, or drum. Small amounts of several other chemicals are added to improve the flotation process, including those described below.

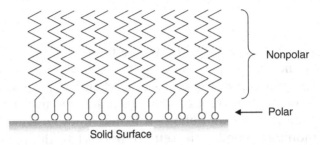

FIGURE 1-14   Monolayer coverage of the amine collector on the surface of a KCl crystal, showing both polar (amine functional group) and nonpolar (long-chain hydrocarbon) parts of the collector molecule. Alignment of collector molecules as shown in this figure results in a hydrophobic layer on the surface of the KCl, but not on the NaCl.

Depressants. Chemicals added to a flotation system to depress (inhibit) the flotation of undesired minerals are known as depressants. In potash, depressants are added to inhibit flotation of any residual insoluble material that was not removed in desliming. In potash flotation, clays are particularly problematic, as they have a very large surface area, so they would adsorb large amounts of the collector. The clays are therefore depressed with inexpensive reagents such as carboxymethylcellulose and guar. These polymeric species are added to the ore slurry before the addition of collector. The polymeric depressants adsorb onto the clay surfaces through hydrogen bonding and coat the surface, thus preventing adsorption of the collector.

Frothers. The flotation process would not work if the air bubbles (with attached particles) rose to the top of the cell and promptly burst—leaving the solids to settle back to the bottom of the cell. Frothers are therefore added to the flotation slurry; they stabilize the air bubbles so that they do not burst when they reach the top of the cell. Instead, the air bubbles form a layer of KCl-enriched foam that is skimmed from the top of the cell. Frothers are usually alcohols that reduce air-water interfacial tension by coating the air bubbles, leaving the polar hydroxyl groups exposed to the solution. In potash, the frothers in use are typically C_6 to C_{10} aliphatic alcohols.

Extenders. Nonpolar chemicals added to the flotation circuit to improve recovery of the desired mineral are known as extenders. The mechanism by which extenders work is unknown. Two proposed theories are that the extender collects near the bubble-solid interface, thus strengthening the bond between them, and that the extender binds to the hydrophobic part of the collector molecules, thus making the surface of the desired mineral more hydrophobic.

Once the ore slurry has been properly conditioned with the collector, and other necessary reagents, it is pumped to the flotation cells, in which the KCl is collected from the top of the cell while the NaCl flows through. In flotation, the ore slurry is agitated in a series of flotation cells, and air bubbles are injected into the mixture at the cell bottom. As the air bubbles rise to the top of the cell, collector-coated KCl particles adhere to the air bubbles and are dragged to the top of the cell, where they are skimmed off into a collection launder.

Each flotation bank is commonly a large rectangular steel tank, subdivided into a number of square cells. Each cell has a large agitator to slurry the ore and provision for the injection of air bubbles. The KCl concentrate froth is collected from the top of each cell by a paddle wheel, which scoops the product from the top of the cell and deposits it into a collection launder. A typical potash flotation circuit is illustrated in Figure 1-15. Flotation concentrate is typically close to the 95% KCl grade that is required for fertilizer, and can be sold after debrining, drying, and sizing.

Product Debrining and Drying

The product obtained from flotation is a slurry, consisting of solids that are a minimum of 95% KCl, and a cosaturated KCl/NaCl brine. The brine must be recovered and recycled for efficient plant operation. The slurry is therefore processed through screen-bowl centrifuges (Fig. 1-16). In a screen-bowl centrifuge the solids are transported past a screen section before they are discharged as a damp cake. The screen section results in a drier cake, and also allows the operators the opportunity to displace some of the KCl/NaCl brine in the product. One of the goals in centrifuging is to reduce the cake moisture as much as possible, because high moisture leads to high drying costs, and because the NaCl in the brine salts dilutes the concentrate grade. The moisture content of the damp cake varies with particle size of the product, and will be in the range of 3–6% moisture. All the brine from the centrifuges is recovered, cleaned (of insolubles) in a thickener, and recirculated to the front of the flotation circuit.

Centrifuge cake is then rigorously dried in cocurrent rotary kiln or fluidized bed dryers. The final product moisture is less than 0.1%, since damp product will cake during storage and shipping. Potash dryers are fired with natural gas.

FIGURE 1-15 Banks of flotation cells used for the production of potash concentrate.

Dryer exhaust gases contain a significant amount of fine particulate, and therefore these off-gases must be cleaned before release to the environment. Many different types of gas cleaning equipment are in use in the potash industry, including wet scrubbers, baghouses, and electrostatic precipitators (Fig. 1-17).

Product Screening and Compaction

The dried product will be on grade (>95% KCl) but with a wide range in particle sizes, varying from 3 mm to dust of less than 30 microns. Such a wide range of sizes is not acceptable to customers. Farmers usually purchase fertilizers that are a blend of all three plant nutrients (N, P, K), and it is important that the sizing of each of the three nutri-

FIGURE 1-16 A set of screen-bowl centrifuges used for debrining potash product. The tank in the central part of the picture is a feed distributor, which accepts the concentrate from flotation and directs a portion of the slurry to each of the centrifuges.

ents is the same; otherwise, segregation of the fertilizer components will occur during storage and transportation, resulting in inhomogeneous distribution of nutrients on the crops.

Potash product is therefore screened into carefully controlled size fractions and sold accordingly. The most common sizes of potash fertilizer are Granular (~1.7–4.0 mm) and Standard (~0.2–1.7 mm). Product screens achieve a separation by allowing the dryer product to flow by gravity over an inclined screen deck. Undersize material is collected from beneath the screen, while oversize material flows off the end of the screen. Modern plants rely on multideck, high-efficiency screens which have high throughput and good screening efficiency (i.e., minimal contamination with off-size material).

FIGURE 1-17 An electrostatic precipitator, which is used to remove fine particulate matter from the exhaust gases from a potash dryer.

After standard and granular products have been collected from the dyer discharge, the plant operator is left with a large portion of under-size potash dust, which is rich in KCl but too small to be sold as product. Producers have two options for dealing with this dust—compaction or crystallization. The compaction process is described in the following paragraphs, while a discussion on crystallization is presented in the next section.

Compaction is a process by which producers convert potash dust into additional granular product by using mechanical roll presses. Dust is fed into the compactor, either by gravity or by a system of mechanical screws known as a "force feeder." Inside the compactor, the dust is compressed under a high pressure of ~12 tons per inch of roll width. Roll sizes in the potash industry vary from 24 to 39 inches, with larger machines becoming the norm. The compactors discharge a sintered "board" that has a high (99.5% of theoretical) density.

The board produced by compactors is very brittle and can easily be crushed in a roll crusher. The crushed board is then screened, and additional granular product is recovered. Oversize from the screens is crushed and rescreened, etc., until all the board is converted to granular product.

Crystallization

There are two principal applications for the crystallization technique in potash processing. In a conventional flotation mill, crystallization is one option for processing the KCl-rich fines collected from the screening plant, as described above. These excess KCl fines can be dissolved in a hot brine to produce a KCl-rich (pregnant) solution, which is then recrystallized to produce a purified (value-added) product. In several plants, however, crystallization is the primary process for production of potash. In the two solution mines, a warm, undersaturated brine is injected into the potash ore zone and pumped back to the surface (some distance away) as a near-saturated brine. The KCl-rich brine is then cooled in a pond or in a series of crystallizers (or both) to produce potash.

The crystallization process utilizes the difference in solubility between KCl and NaCl. In a cosaturated brine (i.e., saturated with respect to both KCl and NaCl) the solubility of KCl in water increases, and the solubility of NaCl decreases, with increasing temperature (Fig. 1-9). Therefore, if a hot (e.g., 100 °C) brine is saturated with both KCl and NaCl and then subsequently allowed to cool (e.g., to 40 °C), it will precipitate KCl from the solution. In practice, the brine is often not cosaturated with both KCl and NaCl; rather, the sodium content of the brine is controlled to somewhat lower levels, and thus the purity of the resulting KCl product can be regulated.

The cooling of hot brines to precipitate KCl is performed in a series of vessels known as crystallizers. A typical crystallizer is shown in Figure 1-18. Crystallizers are sealed vessels in which a vacuum is maintained, causing the surface of the liquor to undergo vigorous boiling. The hot brine is continuously circulated either by an agitator and draft tube (Fig. 1-18) or by an external pump. As a result of the boiling and the constant circulation, the brine undergoes uniform cooling. The KCl crystals that precipitate as a result of the cooling settle to the bottom of the vessel. At the bottom, some crystallizers have an elutriating leg. Brine is circulated into the bottom of the

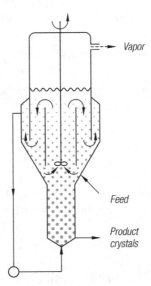

FIGURE 1-18 A draft-tube baffle (DTB) crystallizer used in the potash industry. Variations on this design include the use of a bottom-entry agitator drive and the lack of an elutriating leg.

elutriating leg so that the product crystals are kept fluidized. In such a state, the crystals grow larger and they can also be pumped out of the crystallizer. Other crystallizers simply pump the product slurry from the bottom of the vessel.

The steam that is produced at the top of the crystallizer must be collapsed, or the vigorous boiling (necessary for good cooling) would cease. The steam vapors are generally collapsed in a barometric condenser in which the steam vapors flow countercurrent to a spray of a cooler fluid. Crystallizer circuits are generally arranged in series with the first crystallizer containing the hottest brine; each subsequent stage then cools the brine to a lower temperature. The coolest brine (from the final stage) is then used as the cooling fluid for the second-last crystallizer's condenser (Fig. 1-19). After flowing through this condenser, the cooling fluid is pumped to the third-last condenser, and so on, until the (reheated) cooling fluid exits from the first-stage condenser. Thus, the cooling fluid and the pregnant liquor flow countercurrent to each other through the series of crystallizers. Such an arrangement is useful because it optimizes heat recovery of the crystallizer circuit, since the reheated cooling fluid can then be heated further, resaturated with potash, and recirculated back to the first stage of crystallizers as pregnant liquor.

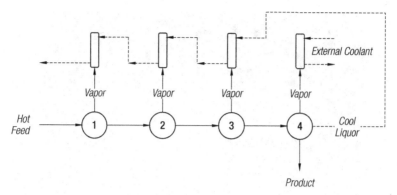

FIGURE 1-19   Typical flows in a four-stage crystallization circuit. Note that the flow of the cooling liquor (------) through the barometric condensers is countercurrent to the flow of the feed (———).

STORAGE, TRANSPORTATION, AND DISTRIBUTION OF POTASH

Production rates from modern potash plants are several hundreds of tons per hour. The demand for potash product varies according to the season and demand from various parts of the world. As a consequence, producers need to install substantial storage and shipping capability.

Storage of finished potash product is generally done in covered buildings up to 200 feet wide and 1600 feet long. Potash, being a salt, will absorb moisture and cake into lumps, much like salt cakes in a salt shaker on a humid day. As a result, it is important that the finished product be protected from rain and snow.

The finished product is transported to storage by conveyor belt, and each product is dispatched into its own storage area by an overhead tripper. On-site storage capacity varies, but is typically in the range of 200,000 tons.

When orders are received, product is reclaimed from storage, usually with a loader or a dozer (Fig. 1-20) that transports the material to a conveyor belt. The conveyor moves the product to the loading facility, where it is screened to remove any remaining dust. The finished product is then conditioned by addition of small quantities of reagents designed to suppress dust and to eliminate caking. Finally, the potash is loaded into trucks or railcars for delivery to customers.

Transportation of 20+ million tons of potash product to customers around the world is a difficult task. A small fraction of the product is

FIGURE 1-20 A storage facility containing potash product. In the foreground is a loader, which is used for moving the product to a reclaim conveyor, which transports the product to the loading facility.

shipped by truck, usually to customers relatively close to the producing plant. The majority of the product, however, is shipped by rail.

Distribution of product within North America is generally achieved by shipping 124-car "unit trains" that carry potash only. The unit trains deliver the product to holding tracks in the Midwest. The unit trains are then separated according to the types of product and customer demand, and the railcars are fanned out to in-market warehouses or potash distribution centers. Individual railcars then fan out to dealers. In-market warehouses vary in size from 1,000 to 60,000 tons and reduce the dealers' need for warehouse space.

Shipment overseas is typically accomplished by delivering unit trains of product to ports. Several different ports are used, with the largest being Vancouver, Canada. Product is received at the port, where it can

FIGURE 1-21   Marine terminal for loading potash product, located at St. John, New Brunswick, Canada.

then be placed into storage. The terminal at Vancouver has storage capacity for 300,000 tons of potash. Storage of such large volumes of potash is a challenge, with many different types of product received from various mines. In some cases, the product is not directed to storage, but rather is directly loaded into waiting vessels (Figs. 1-21 and 1-22). The loading rate for oceangoing vessels is 3000 tons per hour. Once loaded into oceangoing vessels, the finished product is ready for delivery to countries around the world.

POTASH PRODUCTS

Most granular and standard potash is produced for sale as a fertilizer, which must meet a minimum of 95% purity of KCl. Remaining impurities in the final product include low levels of sodium chloride and insoluble minerals such as dolomite, anhydrite, and quartz. The fertilizer-grade material is screened so that it meets exacting standards of particle size, so that it will be easy for the farmer to handle, and so that it can

FIGURE 1-22   An oceangoing vessel being loaded with potash for delivery to off-shore customers.

be blended with other nutrients such as nitrogen and phosphorus. Standard product is shipped predominantly to India and China, while granular material is the preferred product in most other countries. Potash fertilizer is then sold as a crop nutrient throughout the world (Fig. 1-23).

Crystallized products, either from a solution mine or from crystallization of the KCl dust generated from a flotation plant, are typically sold for use as an industrial chemical. This purified KCl is used in the manufacture of KOH for glassmaking, potassium carbonate, and many

FIGURE 1-23   A selection of crops fertilized with potash. Clockwise from the upper left: rice, cotton, bananas, and oil palm.

other chemicals. Some of the industrial-grade KCl is processed through a roll press ("compactor") using equipment similar to that described in the section "Product Screening and Compaction." The compactor fuses the industrial-grade product into large (1 inch) pieces, which can be easily handled and stored without the addition of any conditioning chemicals; such large pieces of purified potash are sold for use in water softening applications. One of the largest North American producers of KCl for water softening is the PotashCorp Cory plant, shown in Figure 1-24.

FIGURE 1-24   PotashCorp Cory Division, which produces fertilizer-grade potash, along with KCl sold for use as a water softening regenerant under the trade name Softouch.

REFERENCE

Brogiotti WE and Horwald FP, inventors; US Patent 972,087. 1975 July 29.

BIBLIOGRAPHY

Banks AF. Milling of potash in Canada. In: Pickett EE, Hall WS, Smith GW, editors. *Milling Practice in Canada*, Special Volume No. 16. Canadian Institute of Mining and Metallurgy; 1985. p 342–348.

Fuzesy LM. Petrology of potash ore in the Middle Devonian prairie evaporite, Saskatchewan. In: McKercher RM, editor. *Potash Technology: Mining, Processing, Maintenance, Transportation, Occupational Health and Safety, Environment*. Proceedings of the 1st International Potash Technology Conference. Pergamon Press; 1983. p 47–57.

Strathdee GG. A global service. *Phosphorus Potassium* 1995; January–February: 34–40.

Strathdee GG. Developing global fertilizer markets. *Phosphorus Potassium* 1996; January–February: 27–35.

Strathdee GC, Haryett CR, Douglas CA, Senior MV, Mitchell J. The Processing of Potash Ore by PCS. 14th International Mineral Processing Congress; 1982; Toronto, Canada. Canadian Institute of Mining and Metallurgy; 1982.

Strathdee GG, Klein MO, Melis LA, editors. Crystallization and Precipitation. Proceedings of the International Symposium; 1987 Oct 5–7; Saskatoon, Saskatchewan, Canada; 1987.

CHAPTER 2

WHAT IS HARD WATER?

DEFINITION OF HARD WATER

The phrase "hard water" originated as a folk term to describe waters, high in calcium and magnesium, that were hard to work with when doing household laundry and cleaning. Calcium and magnesium ions react with soap anions, producing an insoluble soap curd (see the section "Problems associated with hard water"), which reduces the effectiveness of soaps and laundry detergents. As a result, people found that when washing or doing laundry with hard water, it was necessary to add more soap for cleaning to be effective. The hardness of water can then be defined as the content, in solution, of calcium and magnesium, regardless of the type of anions present.

Most of the water used for domestic purposes and consumption in the United States and Canada should be softened by one of the several available methods before it is used. There are categories or degrees of hardness, but before we list these, the origin of hard water and the measurement of water hardness will be discussed.

Water Softening with Potassium Chloride: Process, Health, and Environmental Benefits, by William Wist, Jay H. Lehr, and Rod McEachern
Copyright © 2009 by John Wiley & Sons, Inc.

HOW HARD WATER IS CREATED

Surface waters originate from rain, which has a very low content of dissolved ions (not surprising, given that rain water is distilled by nature!). As rain falls through the atmosphere it absorbs carbon dioxide from the air. It also collects dust particles and leaches out any soluble minerals in these particles. Acid rain is an example of this. Normally when rain water reaches the Earth the total content of dissolved and suspended solids is very low, and therefore the water is very soft. The process of conversion to hard water begins as soon as the rain touches the ground. Once the rain reaches the earth it makes its way into rivers, lakes, and oceans and seeps into the ground and eventually into the ground water table.

As the water flows through the soil, creeks, and rivers toward the sea, it dissolves water-soluble minerals containing calcium, magnesium, chlorides, sulfates, nitrates, carbonates, and bicarbonates and also trace amounts of other elements (see Table 2-1) by reactions illustrated as the following:

$$nH_2O + CaCl_2 \cdot 2H_2O \rightarrow Ca^{2+} + 2Cl^- + (n+2)H_2O$$

$$nH_2O + MgCl_2 \cdot 7H_2O \rightarrow Mg^{2+} + 2Cl^- + (n+7)H_2O$$

The most important minerals involved in the formation of hard water, however, are the carbonates of calcium and magnesium. If it were not for atmospheric carbon dioxide, the dissolution of these carbonate minerals would proceed until a saturated solution was formed, and then would stop. At this point, the total hardness of the water would be quite

TABLE 2-1: Some minerals that contribute to the formation of hard water

Common Name	Chemical Name	Chemical Formula	Solubility
Limestone	Calcium carbonate	$CaCO_3$	15.3 mg/L
Chalk	Calcium carbonate	$CaCO_3$	15.3 mg/L
Marble	Calcium carbonate	$CaCO_3$	15.3 mg/L
Gypsum	Calcium sulfate	$CaSO_4 \cdot 2H_2O$	~2400 mg/L
Epsom salts	Magnesium sulfate	$MgSO_4 \cdot 7H_2O$	~42%
	Calcium chloride	$CaCl_2$	~43%
	Magnesium chloride	$MgCl_2$	~35%
Dolomite	Calcium magnesium carbonate	$CaCO_3 \cdot MgCO_3$	320 mg/L
Magnesite	Magnesium carbonate	$MgCO_3$	106 mg/L

low, for example, a saturated solution of calcium carbonate would contain just $15.3\,mg/L^{-1}$ as $CaCO_3$ (Table 2-1).

Dissolution of limestone continues past this point because of atmospheric CO_2. Carbon dioxide initially enters the water when it is exposed to the atmosphere as rain. Further CO_2 is absorbed as surface and ground waters are exposed to the atmosphere. Along with the carbon dioxide from the air, more carbon dioxide is absorbed from decaying vegetation on and in the ground. The carbon dioxide that is dissolved in the water then produces carbonic acid:

$$H_2O + CO_2 \rightarrow H_2CO_3$$

Carbonic acid is then able to dissolve some of the sparingly soluble minerals like limestone and dolomite by the reactions:

$$H_2CO_3 + CaCO_3 \rightarrow Ca^{2+} + 2HCO_3^-$$
$$2H_2CO_3 + CaCO_3 \cdot MgCO_3 \rightarrow Ca^{2+} + Mg^{2+} + 4HCO_3^-$$

The absorption of atmospheric CO_2 and the concomitant dissolution of limestone and dolomite then continues, producing a dilute solution of the soluble calcium and magnesium bicarbonates. The overall reaction for generation of hard water can then be written, using calcium carbonate (limestone) as a representative source of hardness:

$$H_2O + CO_2 \rightleftharpoons H_2CO_3 + CaCO_3 \rightleftharpoons Ca^{2+} + 2HCO_3^-$$

Absorption of CO_2 from the atmosphere drives the above reaction to the right. Conversion of the CO_2, first to H_2CO_3, and then to HCO_3^-, removes the dissolved CO_2 from the water, so it is able to absorb more CO_2. The process of absorption, reaction, and conversion to calcium bicarbonate is repeated over time until the water becomes hard.

The process of limestone and dolomite dissolution can be illustrated clearly by analysis of river water along a series of points from the headwaters through to the ocean. For example, see the analysis of water from the Colorado River as shown in Table 2-2. The degree of hardness will increase the longer that the water has been in contact with carbon dioxide and calcium and magnesium minerals. Therefore, the degree of hardness is generally lowest in rivers, increases in lake water, and is higher in ground water. Usually deeper wells have much harder water than shallow wells, since the groundwater has taken longer to sink to the deeper levels. Shallow wells in sandy locations tend to have low hardness levels.

TABLE 2-2: Variation in the Total Dissolved Solids (TDS), hardness, and pH for the Colorado River

Location	TDS (mg/L)	Total Hardness mg/L as $CaCO_3$	pH
Hot Sulfur Springs, CO	92	56	7.4
Glenwood Springs, CO	292	159	7.6
Cameo, CO	380	181	7.7
Cisco, UT	530	276	7.7
Lees Ferry, AZ	609	300	7.6
Grand Canyon, AZ	655	307	7.8
Parker Dam, AZ-CA	753	360	7.8

In addition to the dissolution of calcium and magnesium carbonates, carbonic acid also dissolves small amounts of iron and manganese and also trace amounts of other elements. Although not directly related to hardness, these other components do contribute to water quality. The water also can become polluted with industrial, human, and animal waste along with decayed vegetation that can degrade the water quality. Other contaminants include algae and other microorganisms.

PROBLEMS ASSOCIATED WITH HARD WATER

Insoluble Soap Curd Formation

Soaps are typically the sodium salts of long-chain organic acids such as oleic or stearic acid. A typical soap is therefore a compound such as sodium stearate (Na^+ stearate$^-$). In soft water, the sodium stearate molecule will dissolve; the sodium ion does not contribute to cleaning, but the stearate anion is an effective soap. However, in hard water, the calcium and magnesium ions (Ca^{2+} and Mg^{2+}) react with stearate anion to cause an insoluble precipitate known as soap curd or scum. For example:

$$2[Na^+stearate^-] \rightarrow 2Na^+ + 2\,stearate^-$$

$$2\,Stearate^- + Ca^{2+} \rightarrow [Ca^{2+}(stearate^-)_2]\downarrow$$

where the downward arrow indicates the precipitation of an insoluble material. The most familiar and visible soap curd is "bathtub ring." Other problems not so visible but quite significant are deposits on clothes fabrics after they have been washed. Dull whites and colors are caused by soap curd that is not easily removed in the rinse cycles. This also shortens the life of clothes that are washed and worn frequently.

In dishwashers, soap curd causes spotting and streaking of glassware. In the shower, it makes one's skin feel dry and rough unless a lot of soap is used. After rinsing, one's skin still feels dry and scaly and hair feels lifeless and dull.

In addition to the performance problems with soaps in hard water, the precipitation of soap curd causes one to use excess soaps and detergents to compensate for the precipitation of soap curd. Soap, shampoo, and detergent consumption therefore increases with the hardness of water.

Insoluble Scale Deposits

The calcium and magnesium ions along with the bicarbonate ion (HCO_3^-) cause scale formation when hard water is heated. The underlying reason for scale formation is that the solubility of carbon dioxide is reduced as water is heated. As a result, carbonic acid dissociates back into CO_2, which is evolved from the solution:

$$H_2CO_3 \rightarrow H_2O + CO_2 \uparrow$$

where the upward arrow indicates the return of CO_2 from the dissolved to the gaseous state. The reduction in carbonic acid results in the reversal of the reaction by which limestone was dissolved to form hard water, that is:

$$Ca^{2+} + 2HCO_3^- \rightarrow CaCO_3 \downarrow + H_2CO_3 \rightarrow H_2O + CO_2 \uparrow$$

Similarly, for magnesium:

$$Mg^{2+} + 2HCO_3^- \rightarrow MgCO_3 \downarrow + H_2CO_3 \rightarrow H_2O + CO_2 \uparrow$$

The downward arrows in the above two equations indicate the precipitation of $CaCO_3$ and $MgCO_3$ scale. The amount of scale increases as the water is heated, until all available CO_2 has been evolved. Calcium carbonate is the greatest single contributor to scale formation. Inspection of the reaction for $CaCO_3$ scale precipitation shows that it is just the reverse of the reaction for the formation of hard water (see the section "How Hard Water is Created"). It is interesting to consider, then, that the scale that precipitates in your kettle is simply reprecipitation of limestone rocks that were in contact with ground water upstream of your home!

Once precipitated, water will not dissolve this scale. Acidic solutions or other special cleaning chemicals are required to dissolve scale. The scale that precipitates has a very poor ability to transfer heat. Therefore, scale precipitation reduces the efficiency of electric kettles, heat exchangers, water heaters, humidifiers, and electric irons.

Scale formation is not just a problem because of reduced heat transfer efficiency. In addition, hot water lines can become clogged with scale precipitate and flow can be reduced significantly. Even cold water lines, especially at elbows and joints, can become clogged with scale after extended use.

HOW HARD WATER IS MEASURED

Hard water is analyzed by measuring the calcium and magnesium content, which is typically reported as milligrams per liter (mg/L). The results are often reported by calculating the calcium carbonate ($CaCO_3$) equivalent of all the hardness ions in the water. The majority of the hardness ions are calcium and magnesium, but small amounts of other ions can contribute. For simplicity, only calcium and magnesium ions will be considered here.

Conversion to the calcium carbonate equivalent is accomplished by multiplying by the ratio of the equivalent weights (see Chapter 5 for a more detailed explanation of these conversions). For example, the equivalent weight for calcium carbonate is:

$$CaCO_3 = 50.04 \text{ g/eq}$$

while the equivalent weights of calcium and magnesium are:

$$Ca^{2+} = 20.04 \text{ g/eq}$$
$$Mg^{2+} = 12.15 \text{ g/eq}$$

The $CaCO_3$ equivalent of calcium is therefore:

$$Ca^{2+} = \frac{50.04 \text{ g/eq}}{20.04 \text{ g/eq}} = 2.50$$

while the $CaCO_3$ equivalent of magnesium will be:

$$Mg^{2+} = \frac{50.04 \text{ g/eq}}{12.15 \text{ g/eq}} = 4.12$$

TABLE 2-3: Useful conversion factors for water analysis

Unit	Conversion
1 gpg	17.1 mg/L
1 grain	0.064799 g
1 grain	64.8 mg
1 lb	7000 grains
1 lb	453.54 g
1 lb NaCl	5992 grains $CaCO_3$
1 lb KCl	4697 grains $CaCO_3$

The hardness units used in the water softening industry are grains per U.S. gallon, abbreviated as gpg. Most water analysis is reported as milligrams per liter (mg/L) and then converted to gpg. Table 2-3 provides a summary of useful conversion factors.

Water treatment specialists prefer to receive water analysis reports in $CaCO_3$ equivalents in gpg. As an example, if a sample of water is analyzed and contains 130 mg/L calcium and 75 mg/L magnesium, then the total hardness of this sample of water would be:

$$\frac{130}{17.1} \times 2.50 + \frac{75}{17.1} \times 4.12 = 37.1 \text{ gpg as } CaCO_3$$

The equivalent weights as well as the $CaCO_3$ equivalent are given in Table 2-4 for many common components of hard water.

UNIFORM DEGREES OF HARDNESS

Water can be labeled as hard or soft, depending on the calcium and magnesium content. For uniformity, the water softening industry uses the guidelines shown in Table 2-5 to describe the hardness or softness of water.

TYPES OF HARDNESS

There are two types of hardness, temporary and permanent. Temporary hardness is the measure of calcium and magnesium bicarbonates, $Ca(HCO_3)_2$ and $Mg(HCO_3)_2$ dissolved in water, expressed as calcium carbonate. It is called temporary hardness because these bicarbonates

TABLE 2-4: Common constituents of hard water, with their corresponding equivalent weights and CaCO₃ equivalents

Constituent	Equivalent Weight	CaCO$_3$ Equivalent
$CaCO_3$	50.04	1.00
Ca^{2+}	20.04	2.50
Mg^{2+}	12.15	4.12
$CaSO_4$	68.05	0.735
$MgSO_4$	60.18	0.832
$CaCl_2$	55.49	0.902
$MgCl_2$	47.61	1.05
$Ca(HCO_3)_2$	81.06	0.617
$Mg(HCO_3)_2$	73.14	0.684
K^+	39.09	1.28
Na^+	22.99	2.18
KCl	74.55	0.671
$NaCl$	58.44	0.856
Cl^-	35.45	1.41
SO_4^{2-}	48.03	1.04
HCO_3^-	61.02	0.820
CO_3^{2-}	30.00	1.67

TABLE 2-5: Standard terms used to describe the hardness of water

Term	Grains/Gallon (gpg)	mg/L as CaCO$_3$
Soft	<1.0	<17.1
Slightly Hard	1.0 to 3.5	17.1 to 60
Moderate Hard	3.5 to 7.0	60 to 120
Hard	7.0 to 10.5	120 to 180
Very Hard	>10.5	>180

break down under elevated temperatures (such as in a domestic hot water heater) and form scale, according to the reactions:

$$Ca^{2+} + 2HCO_3^- \rightarrow CaCO_3\downarrow + H_2CO_3 \rightarrow H_2O + CO_2\uparrow$$

$$Mg^{2+} + 2HCO_3^- \rightarrow MgCO_3\downarrow + H_2CO_3 \rightarrow H_2O + CO_2\uparrow$$

Permanent hardness is due to the presence of non-bicarbonate ions such as chlorides and sulfates of calcium and magnesium. These minerals do not form scale with the addition of heat because they do not undergo decomposition reactions the way bicarbonate does. When hard water is heated, scale precipitation will stop when all of the bicarbonate has decomposed to CO_2 and been evolved. Any calcium ions

TABLE 2-6: Sample calculations for determination of total, temporary, and permanent hardness

Water Analysis	Sample 1		Sample 2	
Constituent	mg/L	gpg as $CaCO_3$	mg/L	gpg as $CaCO_3$
Ca	67	9.8	62	9.1
Mg	28	6.7	25	6.0
Na	46	5.9	124	15.8
K	33	2.5	31	2.3
Total cations as $CaCO_3$		24.9		33.2
Cl	79	6.5	70	2.8
SO_4	74	4.5	94	5.7
HCO_3	288	13.8	452	21.7
Total anions as $CaCO_3$		24.8		33.2
Total hardness	283	16.5	258	15.1
Temporary hardness	236	13.8	371	21.7
Permanent hardness	47	2.7	−113	−6.6

remaining in solution at that time will be associated with other anions (rather than bicarbonate).

Total hardness is the sum of all hardness constituents (both permanent and temporary) in water expressed as calcium carbonate.

To calculate temporary and permanent hardness, all cations and anions in the water should be analyzed. The total hardness is calculated using the calcium and magnesium ions. Temporary hardness is calculated using the bicarbonate ion. If the temporary hardness exceeds the total hardness, then it is assumed that all the hardness is temporary hardness and there is no permanent hardness. Two examples of these calculations are given in Table 2-6.

Sample 1 in Table 2-6 has a total hardness of 16.5 gpg as calcium carbonate, which is the sum of the calcium and magnesium hardness (9.8 + 6.7). The temporary hardness is 13.8, equal to the bicarbonate hardness. The permanent hardness is then the difference between the total and temporary hardness (16.5 − 13.8 = 2.7 gpg).

Total and temporary hardness for Sample 2 are calculated the same as for Sample 1, described above. However, we note that the calculated permanent hardness is −6.6. In such a case, one would assume that the permanent hardness was equal to zero, and all hardness was temporary.

CHAPTER 3

LOWERING WATER HARDNESS

There are many different processes used to lower the hardness level in water; some of the more widely used in residential and industrial applications include:

- Ion exchange
- Deionization (demineralization)
- Reverse osmosis
- Distillation
- Precipitation

Reverse osmosis, deionization, precipitation, and distillation are processes that reduce water hardness by removing minerals from the water. The total dissolved solids in the water are significantly reduced or eliminated by these processes. In contrast, ion exchange reduces water hardness by removing hardness ions such as calcium and magnesium and replacing them with more desirable ions such as sodium and potassium. However, in ion exchange, the total dissolved solids in the water remain basically the same.

Water Softening with Potassium Chloride: Process, Health, and Environmental Benefits,
by William Wist, Jay H. Lehr, and Rod McEachern
Copyright © 2009 by John Wiley & Sons, Inc.

ION EXCHANGE

Water softening (i.e., removal of Ca^{2+} and Mg^{2+}) by cation exchange is used worldwide. In this process, raw water is passed through a container that holds ion exchange resin beads. The beads are made of a polymer, to which charged functional groups are attached. The functional groups attract and loosely bind counterions of opposite charge (to ensure charge neutrality). The counterions can be transferred from the resin to the solution, in return for other types of ions, that is, the ions can be *exchanged*. In household softeners, for example, the polymer in the resin has charged $-SO_3^-$ functional groups. Cations are attracted to the negative $-SO_3^-$ groups to neutralize the electrical charge. Calcium removal from hard water then occurs by the reaction:

The above reaction is shifted to the right while hard water passes through the ion exchange column and the calcium and magnesium ions in solution are replaced with sodium. The reaction shifts to the right because the resin has an inherent affinity for calcium and magnesium, in preference to sodium. Note that the attachment of one divalent calcium ion to the resin requires the removal of two monovalent sodium ions from the resin (to preserve electrical charge neutrality).

For ease of notation we replace the stylized polymer with "R" and rewrite the equation as:

$$2R-SO_3^-Na^+ + Ca^{2+} \rightleftharpoons (R-SO_3^-)_2 Ca^{2+} + 2Na^+$$

When an adequate volume of hard water has passed through the bed of ion exchange resin, the beads will become saturated with calcium and magnesium; further water that passes through the resin will not be softened. At such time the column is regenerated by passing a strong solution of sodium chloride through it. The excess of sodium shifts the ion exchange equation (above) to the left, and the ion exchange resin

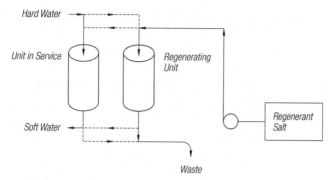

FIGURE 3-1   Schematic of a typical industrial apparatus for water softening.

in the column is reloaded with sodium. Excess amounts of sodium are required to recharge the resin to overcome the inherent affinity of the resin for calcium and magnesium.

A typical industrial softening circuit will be designed with two separate cation exchange columns, so that one unit can be online while the other is being regenerated, as shown in Figure 3-1.

In contrast, most home softeners have only one tank of ion exchange resin. When the resin requires regeneration the softener is bypassed, and the household consumes hard water briefly, while the regeneration occurs.

DEIONIZATION OR DEMINERALIZATION

Deionization is the removal of all ions from solution by a two-step ion exchange procedure. This process uses both a cation and an anion resin. The cation resin removes all the positively charged ions such as Ca^{2+}, Mg^{2+}, K^+, and Na^+ and replaces them with hydrogen ion (H^+). The anion resin removes all the negatively charged ions such as Cl^-, SO_4^{2-}, and HCO_3^- and replaces them with hydroxide (OH^-). The hydrogen and hydroxide ions then react to form pure water. When the resins have been depleted they must be recharged. The cationic resin is recharged with a strong acid, usually hydrochloric also known as muriatic acid (HCl), while the anionic resin is regenerated with sodium hydroxide, better known as caustic soda (NaOH).

Typical reactions for water deionization can be illustrated, for water containing the ions Ca^{2+}, K^+, Cl^-, and SO_4^{2-}. In the service cycle the cationic ion exchange reactions would be:

$$2R-SO_3^-H^+ + Ca^{2+} \rightarrow (R-SO_3^-)_2 Ca^{2+} + 2H^+$$

$$2R-SO_3^-H^+ + 2K^+ \rightarrow 2R-SO_3^-K^+ + 2H^+$$

while in the anion exchange resin, anions are replaced with hydroxide ion:

$$R-OH^- + Cl^- \rightarrow R-Cl^- + OH^-$$

$$2R-OH^- + SO_4^{2-} \rightarrow R_2-SO_4^{2-} + 2OH^-$$

The hydrogen ions produced by the cation exchange resin will promptly react with the hydroxide ions produced by the anion exchange resin to produce water:

$$H^+ + OH^- \rightarrow H_2O$$

In the regeneration cycle, the cation and anion exchange resins are returned to the H^+ and OH^- form, respectively, by the reactions:

$$(R-SO_3^-)_2 Ca^{2+} + 2H^+ \rightarrow 2R-SO_3^-H^+ + Ca^{2+}$$

$$R-Cl^- + OH^- \rightarrow R-OH^- + Cl^-$$

In the regeneration cycle, the above reactions are forced to the right, that is, the resins are returned to the H^+ and OH^- states, by using strong acids and bases such as hydrochloric acid and sodium hydroxide. Regeneration requires large amounts (i.e., an excess) of the strong acid and base, so the cost to the consumer, and the environment, can be significant.

For household users, small, throwaway cartridges are available for limited quantities of drinking and cooking water. The cartridges have a mixed bed of cation and anion resins. If the dissolved minerals in the potable water are high, then the life of the cartridges is reduced significantly. These cartridges tend to be quite expensive.

Large business or industrial applications often use separate exchange tanks for water demineralization. Either the resin can be removed from the tanks and regenerated in special tanks of regenerant or the regenerant can be passed through the exchange tanks. A schematic of a fairly typical industrial water demineralization apparatus is illustrated in Figure 3-2. In this example, raw water passes through a cation exchange column, and then reports to a separate column containing anion exchange resin.

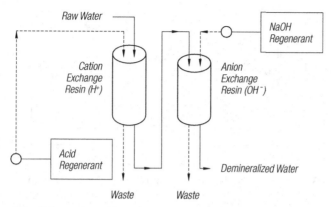

FIGURE 3-2 Schematic of a typical industrial apparatus for water demineralization.

Many businesses use portable exchange tanks. When the resin has been depleted, a recharged tank is installed and the depleted tank is sent to a central location for regeneration. These portable exchange tanks will produce significant quantities of very pure water.

There are definitely some disadvantages of deionization. Throwaway cartridges and regeneration are expensive. Extreme caution must be used with the regenerant reagents such as sodium or potassium hydroxide and hydrochloric or sulfuric acid. These reagents are very corrosive and can be extremely dangerous to handle. As a consequence, demineralization tends to be more common in large industrial applications such as boiler water treatment, and less common in household use.

REVERSE OSMOSIS

Osmosis is the normal flow of a solvent through a semipermeable membrane, in which the solvent flows from a dilute solution on one side to a concentrated solution on the other. The flow will stop when the concentration of solvent on both sides of the membrane is equal, or when the osmotic pressure prevents any further movement of the solvent.

The movement of water from soils into plant roots is an example of osmosis. The regulation of fluids within the cells of the body is also governed by osmosis, which explains why drinking seawater can be fatal. Immediately after drinking seawater, the osmotic process draws water from the body's cells, which will eventually dehydrate them and can lead to death.

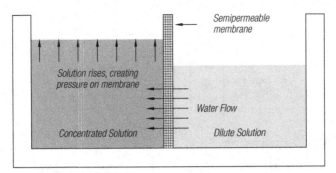

FIGURE 3-3   Flow of water through a semipermeable membrane due to osmosis.

The process of osmosis is also used in clothing. Gore-Tex cloth has millions of small holes that allow water vapor from the body to pass through but will not allow liquids to pass.

A simple diagram illustrating the principles of osmosis is given in Figure 3-3.

When the concentration of both solutions is equal, the water flow will stop. Alternately, when the pressure on the membrane due to the rise of the liquid level on the side with concentrated solution equals the osmotic pressure of the membrane, the water flow will stop.

Reverse osmosis (RO) is a process that reverses, by the application of pressure, the flow of water through a semipermeable membrane. In reverse osmosis, water passes from the more concentrated solution to the more dilute solution through a semipermeable membrane. The membrane will allow only the water molecules to pass through, retaining the highly concentrated dissolved species (ions) on the opposite side of the membrane. The applied pressure must exceed the osmotic pressure of the membrane for reverse osmosis to work. Reverse osmosis is also known as hyper-, nano-, and ultrafiltration. Reverse osmosis is illustrated in Figure 3-4.

Reverse osmosis will remove most particles larger than 0.0005 microns. These include most dissolved minerals including the common ions such as chlorides, sulfates, sodium, potassium, calcium, and magnesium along with other undesirables such as sediment, bacteria, pyrogens, organic contaminants, viruses, cryptosporidium, fluorides, asbestos, as well as most heavy metals. The membrane will allow only the water molecule to pass through, retaining the highly concentrated salts and solids on the opposite side of the membrane.

There are many different sizes of RO units. For the household consumer there are many under-sink devices using line water pressure for

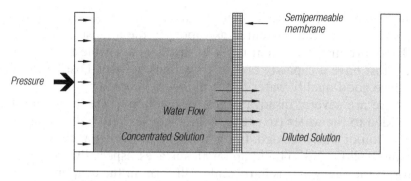

FIGURE 3-4 Water purification by reverse osmosis.

producing drinking water at a rate up to 15 gallons per day. There are also countertop units available that have no storage tank of their own but produce water directly into a container. They can be connected to a faucet at night, and the unit can fill a 5-gallon container by morning.

There are some disadvantages of home RO systems. RO systems require sediment and carbon filtration units to prevent membrane fouling. Water with over 3 gpg hardness should be softened before passing through the RO unit; otherwise, calcium deposits will quickly make the membrane less permeable. The process is generally slow and can be quite inefficient. For example, it may require from 3 to 15 gallons of untreated water to produce 1 gallon of purified water. RO is ideal for supplying good-quality drinking water but can become expensive for providing all the household water required. If the potable water supply is unfit even for bathing and cooking, then RO is definitely an alternative.

DISTILLATION

Distillation has been known for thousands of years, and is widely used as a means of chemical separation and purification. Industrially, distillation remains the workhorse of chemical engineering separation techniques and is widely used in the petroleum and chemical industries. In water treatment, distillation is used for the desalination of seawater, boiler water treatment, as well as for the production of purified potable water.

In the distillation process, water is heated in an evaporator until it boils, producing a steam vapor. The steam vapor is then condensed and collected. Most of the dissolved solids and liquids with higher boiling points than water remain behind. Liquids that boil at lower temperatures than water can be vented as vapors to the atmosphere.

There are many different types of distillers available for household use. Most have a capacity of 3 to 15 gallons per day. Distillation does produce good-quality water for drinking and cooking.

There are several disadvantages to distillation. The heat that must be added to the water (to cause boiling) causes very high energy costs per gallon of water treated. In addition, high maintenance is required, because solids tend to build up on all surfaces, especially in the evaporator. If the raw water is hard, scale will form in the evaporator, and it is generally very difficult to remove. An acid is required to remove this scale during periodic cleaning. If the water hardness is greater than 3 gpg, then the water should be presoftened.

PRECIPITATION

Water purification by precipitation is not common in household applications. However, it is important industrially as well as in the large-scale production of potable water by cities and municipalities. Therefore, a few words about this process are warranted.

There are two main types of precipitation reactions that are used in water treatment: precipitation by addition of lime and by addition of soda ash (sodium carbonate).

Lime softening is widely used to reduce the calcium content of process water, especially when large volumes of water need to be treated. For example, cities often use this method to produce 10^{10} to 10^{12} liters of potable water per year. The addition of lime (calcium oxide, CaO) to water as a means of reducing water hardness (calcium and magnesium) seems initially to be counterintuitive. However, when one examines the underlying chemistry the reason why the process works becomes clear. Lime is added to the water and dissolves to yield two hydroxide ions per molecule of lime:

$$CaO + H_2O \rightarrow Ca^{2+} + 2OH^-$$

The hydroxide ions then convert bicarbonate ions to carbonate:

$$2OH^- + 2HCO_3^- \rightarrow 2H_2O + 2CO_3^{2-}$$

The net effect of lime dissolution is thus:

$$CaO + 2HCO_3^- \rightarrow Ca^{2+} + H_2O + 2CO_3^{2-}$$

The calcium ion liberated on the right-hand side of this equation will promptly precipitate out with a carbonate ion, so we can write an overall reaction:

$$CaO + 2HCO_3^- \rightarrow CaCO_3 + H_2O + CO_3^{2-}$$

The excess carbonate ion on the right-hand side of this equation is then able to precipitate additional calcium and magnesium ions. Note that the overall reaction for lime softening produces one mole of carbonate ions per mole of lime added to the solution.

Reaction of the excess carbonate will then lead to the precipitation of more calcium as calcium carbonate:

$$Ca^{2+} + CO_3^{2-} \rightarrow CaCO_3$$

The overall reaction for the precipitation of calcium and magnesium by the addition of lime will therefore be:

$$Ca(HCO_3)_2 + CaO \rightarrow 2CaCO_3\downarrow + H_2O$$
$$Mg(HCO_3)_2 + CaO \rightarrow CaCO_3\downarrow + MgCO_3\downarrow + H_2O$$

In addition to the above reactions, the fact that lime is a base means that the addition of lime will result in an increase of the alkalinity (pH) of the water, which will often lead to the precipitation of further magnesium as the insoluble hydroxide:

$$MgCO_3 + Ca(OH)_2 \rightarrow CaCO_3\downarrow + Mg(OH)_2\downarrow$$

The calcium and magnesium carbonate, as well as any magnesium hydroxide that precipitates, can be removed quite easily from the potable water by allowing the precipitated solids to settle out in a piece of equipment known as a clarifier. The purified water that overflows from the clarifier will have reduced calcium and magnesium hardness.

Many industrial and municipal water treatment plants use lime softening to reduce the total hardness of the raw water because lime is a relatively cheap chemical. In addition, lime softening is proven technology, and the operation of a lime softening process is relatively easy.

As discussed above, lime softening is a widely used process in which the calcium and magnesium content of hard water is reduced by the addition of lime. Ultimately, the effectiveness of lime softening depends on the availability of bicarbonate ion for the reaction:

$$2OH^- + 2HCO_3^- \rightarrow 2H_2O + 2CO_3^{2-}$$

As a result, lime softening will not be effective for treatment of water that has little or no bicarbonate ion. Furthermore, in lime softening, complete removal of calcium and magnesium cannot be achieved, because once the bicarbonate has all reacted no further softening will occur. Because of these limitations, some industries and cities use soda softening to reduce water hardness. In soda softening, sodium carbonate is added to the hard water. The sodium carbonate will dissolve:

$$Na_2CO_3 \rightarrow Na^+ + CO_3^{2-}$$

The carbonate ions will then be readily available to precipitate calcium and magnesium according to the precipitation reactions:

$$Ca^{2+} + CO_3^{2-} \rightarrow CaCO_3$$
$$Mg^{2+} + CO_3^{2-} \rightarrow MgCO_3$$

Soda softening is not as widely used as lime softening because sodium carbonate is more expensive than lime. However, in some cases it is used where the hard water is low in bicarbonate. Alternately, soda softening may be used as a final (polishing) softening stage after lime softening, as a way to reduce the water hardness lower than lime softening alone can achieve.

BIBLIOGRAPHY

Barnes D, Wilson F. *Chemistry and Unit Operations in Water Treatment.* Applied Science Publishers; 1983.

Brault JL, Degrement A. *Water Treatment Handbook.* Volume 1: 6th Edition; 1991.

Cheremisinoff PN. *Handbook of Water and Wastewater Treatment Technology.* Marcel Dekker; 1995.

Cheremisinoff NP, Cheremisinoff PN. *Water Treatment and Waste Recovery.* PTR Prentice Hall; 1993.

Hopkins ES, Bean EL. *Water Purification Control.* 4th Edition. Williams & Wilkins; 1966.

Introduction to Water Treatment. Volume 2. American Water Works Association; 1984.

James GV. *Water Treatment.* 4th Edition. Technical Press; 1971.

Nordell E. *Water Treatment for Industrial and Other Uses*. Reinhold Publishing; 1951.

Sincero AP, Sincero GA. *Physical–Chemical Treatment of Water and Wastewater*. IWA Publishing: CRC Press; 2003.

Solt GS, Shirley CB. *An Engineers' Guide to Water Treatment*. Avebury Technical Academic Publishing Group; 1991.

Water Treatment Plant Design. 3rd Edition. American Water Works Association; American Society of Civil Engineers; 1990.

CHAPTER 4

THE ION EXCHANGE PROCESS

Ion exchange processes are common in industry today because of the ability of ion exchange resins to selectively remove undesired ions from solution and replace them with an equivalent amount of a different ion. The functional groups in modern ion exchange media are generally attached to a polymer that is robust enough to survive numerous ion exchange cycles over an extended period of time. The polymer is generally shaped into small spherical beads referred to as ion exchange resins.

The earliest ion exchange media used were not based on synthetic polymers, but rather they were modifications of the naturally occurring ion exchange materials found in nature. Clays and other aluminosilicates found in soils are capable of ion exchange. Thompson and Way reported in 1850 that ammonium and calcium ions could be exchanged when solutions were passed through soil samples. Further studies later showed that the process was reversible, and by 1905 a commercial process for water softening was developed that used synthetic sodium aluminosilicate cation exchangers (zeolites). In the first half of the twentieth century several other types of ion exchangers were developed including sulfonated coal and functionalized polymers based on condensation of phenol with formaldehyde. In 1944 D'Alelio developed cation exchange resins based on the copolymerization of styrene and divinylbenzene; the resulting ion exchange resins have proven to work exceptionally well and are now the most widely used polymers for the ion exchange process.

SYNTHESIS AND STRUCTURE OF ION EXCHANGE RESINS

The properties of polymeric ion exchange resins are a result of the presence of functional groups attached to the polymer. The functional groups are usually charged, and as a result an ion pair is formed in order to maintain electrical neutrality. The sulfonic acid functional group, which is common in cation-exchange resins, is illustrated in Figure 4-1. The counterion in the ion pair is only weakly bonded to the functional group and is therefore exchangeable.

The most common type of polymer used in ion exchange resins is a styrene divinylbenzene (DVB) copolymer. This polymer is formed by mixing styrene and DVB in the appropriate amounts, along with an equal amount of water, in a liquid reactor. A surfactant is added, and the mixture is agitated so that the organic liquid becomes dispersed in the water, with an average droplet size of about 1 millimeter. A polymerization initiator is then added, and the resulting polymerization reaction results in the formation of styrene/divinylbenzene beads.

Functional groups are added to the ion exchange beads after the polymerization reaction has been completed. To produce a cation exchange resin with sulfonic acid functional groups, the beads are reacted with concentrated sulfuric acid. The sulfonic acid functional groups thus formed are attached to the polymer throughout the bead and remain in the ionic state, thus leading to the formation of ion pairs.

To make anion exchange resins, the attachment of functional groups is also performed after the styrene/divinylbenzene beads have been formed. The reaction is performed in two steps, with the first step being

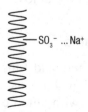

FIGURE 4-1   Typical structure of a functional group and ion pair on a cation exchange resin.

the attachment of chloromethyl groups to the benzene rings, and subsequent amination to produce quaternary amine functional groups.

Functional requirements on ion exchange resins are rather severe. They must be insoluble, must have a high loading capacity, and must be resistant to fracture. High loading capacity is achieved because the reactions attaching functional groups to the polymer occur throughout the bead. As a result, the ion exchange process does not just occur on the surface of the bead; rather, diffusion and exchange of ions occurs throughout the bead (99% of the capacity is inside the bead).

Ion exchange resins will typically expand and contract as they go through the exhaustion and regeneration cycles of the process, and the beads are therefore inherently prone to fracture and breakage. Divinylbenzene is copolymerized with styrene to impart strength to the resulting bead due to the cross-linking that DVB provides. Resins rich in DVB are stronger, but will have a restricted ability to swell when wetted, which can lower the capacity of the resin. Highly cross-linked resins can also have impaired functioning because of the slower kinetics associated with diffusion of ions through a highly cross-linked structure.

The final ion exchange resin bead will be spherical, with ample functional groups that will attract exchangeable counterions. The functional groups are not restricted to the surface of the bead; ion exchange resins are designed to be porous enough that counterions can diffuse through the bead to functional groups on the inside. The size of the final resin bead product will correspond approximately to the size of the initial styrene/divinylbenzene droplets before polymerization; bead size is typically about 1 mm, but can range from 0.2 to 1.7 mm.

TYPES OF ION EXCHANGE RESINS

There are four types of ion exchange resins in common usage. They are strong and weak acid cation exchange resins, as well as strong and weak base anion exchange resins. The structure and properties of each type of resin are described briefly below.

a) Strong Acid Cation Resins

Strong acid resins have sulfonic acid functional groups. They are the most commonly used ion exchange resins, and are widely used in home softeners and (in conjunction with anion exchange resins) in demineralization processes. Advantages of this type of resin include a relatively modest cost, resistance to breakage, and nonselective removal of all cations. However, this type of resin will require a substantial amount of regenerant, especially if it is being regenerated with an acid.

b) Weak Acid Cation Resins

The exchange site on weak acid cation resins is the carboxylic acid functional group. Although not as commonly used as strong acid cation exchange resins, these resins can have specialized applications because of improved selectivity for Ca^{2+} and Mg^{2+}. In addition, these resins can be efficiently regenerated with acids so they can be used in a two-bed system along with strong acid resins. In such a case the weak acid resin can be regenerated with the spent acid from regeneration of the strong acid resin, thus improving process efficiency and economics.

c) Strong Base Anion Resins

The exchange site on strong base anion exchange resins is a quaternary ammonium functional group. The two common quaternary ammonium groups are Type 1, with three methyl groups attached to the nitrogen, and Type 2, which has two methyl groups and an ethanol group. Both are effective in removing all types of anions from water, although the Type 2 resins can be less effective at removing silica and carbon dioxide.

Strong base anion exchange resins are commonly used, along with strong acid cation exchange resins, in demineralization. Like the strong acid cation resins, regeneration of strong base anion exchange resins is generally inefficient and will require the use of excess regenerant.

d) Weak Base Anion Resins

The exchange site in weak base anion exchange resins is an amine functional group. These resins can be used for deacidification as well as for the selective removal of sulfates or chlorides. In addition, they can be efficiently regenerated with near-stoichiometric quantities of a strong base.

HOUSEHOLD WATER SOFTENING

Household water softeners contain strong acid ion exchange resins. The sulfonic acid functional groups attract cationic counterions to maintain electrical charge neutrality, as shown in Figure 4-1. The counterions are exchangeable, so that the sodium or potassium loaded onto the resin (during the regeneration cycle) will be released by the resin when calcium ions in hard water become attached. The reaction that will occur during softening will be as described in Chapter 3 in the section on ion exchange:

During the service cycle, hard water will be softened and the above reaction will shift to the right. During regeneration, the brine solution (in this case, with sodium chloride) will shift the reaction back to the left so that the softening process can be repeated.

Ion exchange resin beads are attracted to some ions more than others (for a detailed discussion on this subject, see the section "Mathematical Treatment of Ion Exchange Equilibria"). For a typical cation exchange resin, the order of preference for attachment to the beads in a dilute solution is:

$$Ca^{2+} > Mg^{2+} > K^+ > Na^+$$

Therefore, during the service cycle, calcium and magnesium ions will readily load onto the resin beads. Regeneration is more difficult, and an excess of sodium or potassium chloride must be used in a strong solution to force the ion exchange resin back into the Na^+ or K^+ state.

Symbolically, we can represent an ion exchange bead when fully regenerated, with KCl, as shown in Figure 4-2. Similarly, an ion exchange bead late in the service cycle, and nearly fully loaded with calcium and magnesium, can be represented as in Figure 4-3. We see from Figure 4-3 that the resin beads are not fully converted to the calcium and magnesium form—a small number of potassium ions remain attached. This illustrates the reality that since ion exchange is a chemical equilibrium process, conversion from one state to the other will not proceed completely unless a large excess of reagent is used. In practice, household softeners are set to regenerate approximately 90%, which provides good softening and cost-effective use of regenerant.

The underlying chemical reaction during household water softening will be the replacement of calcium and magnesium ions with sodium

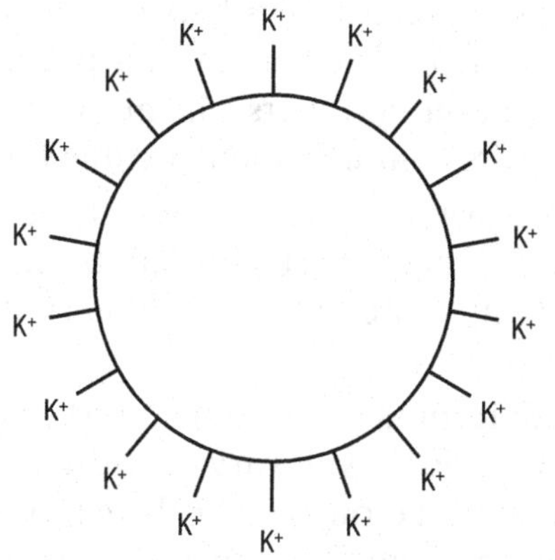

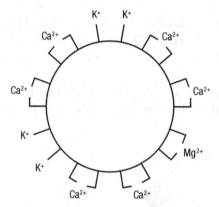

FIGURE 4-2 Representation of a cation exchange resin bead fully loaded with potassium ions.

FIGURE 4-3 Representation of a cation exchange resin bead late in the service cycle, when nearly fully loaded with calcium and magnesium ions.

or potassium. As a result, household softeners do not reduce the total content of dissolved solids in the water. Instead, household softeners simply remove the troublesome calcium and magnesium ions, and replace them with ions that do not cause problems with soap performance or scale buildup.

TYPICAL HOUSEHOLD WATER SOFTENERS

There are some variations in the design of household water softeners, but generally they consist of the following components:

- A pressure vessel, which contains between 0.5 and 2.0 cubic feet of cation exchange resin beads. Generally, strong acid cation exchange resins are used in home softeners. The most common size of household water softeners contains about 1 cubic foot of beads.
- A salt/brine storage tank, which contains the sodium chloride or potassium chloride regenerant salt. The brine solution for regeneration is mixed and stored in this tank until it is required for the regeneration cycle.
- A device that determines when the next regeneration cycle is required. These devices will vary in design, and can trigger a regeneration cycle based on time or the gallons of water softened. Alternately, they can contain a sensing element that measures the residual hardness in the treated water and initiates a regeneration cycle when the quality of the treated water starts to degrade.

At the start of the service cycle, the ion exchange beads will be predominantly in the potassium or sodium form. Hard water will flow through the pressure vessel, and soft water will be discharged, as shown in Figure 4-4, for the case of an ion exchange resin initially loaded with potassium ions.

As the service cycle continues, the beads will be stripped of potassium and will start to become loaded with calcium and magnesium.

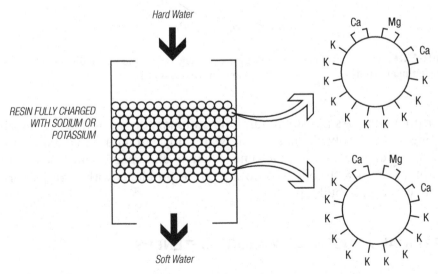

FIGURE 4-4 Bed of ion exchange resin, initially loaded with K⁺, at the start of the service cycle.

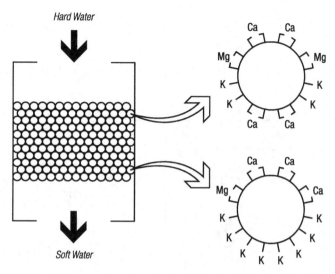

FIGURE 4-5 Bed of ion exchange resin, initially loaded with K$^+$, partway through the service cycle.

Initially, the beads at the top of the column, exposed to the hardest water, will become loaded predominantly with calcium and magnesium first, as shown in Figure 4-5.

Later, the ion exchange resin near the bottom will become loaded with calcium and magnesium. When the beads near the bottom of the column are substantially loaded with calcium and magnesium, the ability of the unit to produce high-quality soft water will become compromised, and a regeneration cycle must be initiated. At this point the condition of the resin bed will be as shown in Figure 4-6.

When the regeneration cycle begins, the softening unit will be removed from service (hard water will not flow through the unit), and instead salt brine (sodium chloride or potassium chloride) will flow through the ion exchange resin bed. Figures 4-7 through 4-9 show the conversion of the resin back predominantly into the K$^+$ form. Note that the regeneration does not typically restore the resin completely into the K$^+$ form, as it is more efficient to regenerate only until the beads are predominantly in the K$^+$ form.

Once the resin has been restored to being predominantly in the K$^+$ state, then the unit will be returned to the service cycle, and processing of hard water will continue.

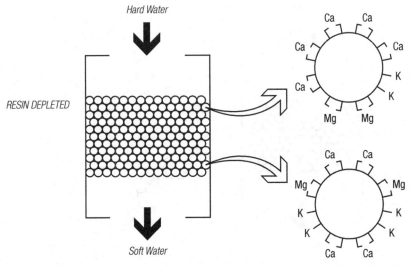

FIGURE 4-6 Bed of ion exchange resin, initially loaded with K^+, after processing many gallons of hard water, and at the end of the service cycle. At this point in the cycle, small but significant amounts of hardness will begin to report to the effluent (the "soft" water).

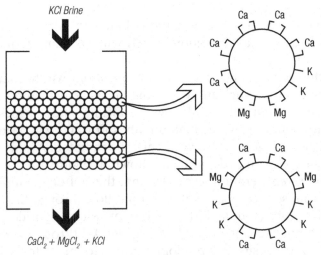

FIGURE 4-7 Bed of ion exchange resin at the beginning of the regeneration cycle. In this illustration, the resin is being regenerated by a KCl brine, to restore the resin to the K^+ form. In this example, there is cocurrent, or downward flow of the brine.

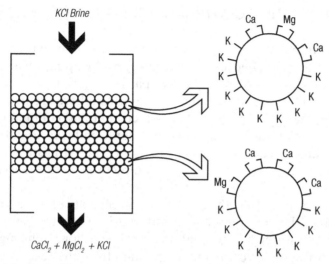

FIGURE 4-8   Bed of ion exchange resin nearing the end of the regeneration cycle. In this illustration, the resin is being regenerated by a KCl brine, to restore the resin to the K⁺ form. The flow of brine is, in this example, downward, or cocurrent (with the water flow).

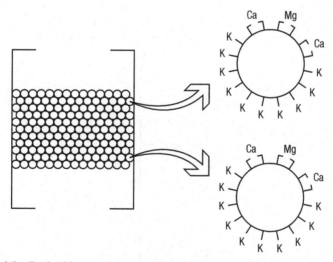

FIGURE 4-9   Bed of ion exchange resin at the end of the regeneration cycle and ready to be returned to service. Note that at this point the resin has not been converted 100% back into the K⁺ state because complete regeneration requires the use of a large excess of KCl brine, which is inefficient use of the regenerant.

COCURRENT AND COUNTERCURRENT REGENERATION

All conventional water softener systems utilize the expandable bed principle. The resin bed expands during backwashing, which allows contaminants and any precipitated particles (such as iron) to be removed with the backwash stream. In contrast, there are two methods of injecting the regenerant solution—cocurrent and countercurrent.

In the cocurrent flow system, the regeneration brine solution (NaCl or KCl) is introduced to the top of the unit. Then it passes downward through the resin bed in the same direction as the service flow (Fig. 4-10).

Two factors may affect the operation and capacity of a cocurrent flow system. First, the regenerant is introduced into the fresh water area, displacing the water and diluting the brine solution, reducing its efficiency (at stripping calcium and magnesium from the resin) during the beginning of the regeneration cycle. At very low regenerant dosages, not all of the resin in the bed may be fully regenerated; this leaves a layer only partially regenerated, resulting in hardness leakage when the unit is returned to service. Second, in cocurrent regeneration, the ion exchange resin at the bottom of the column will not be regenerated as

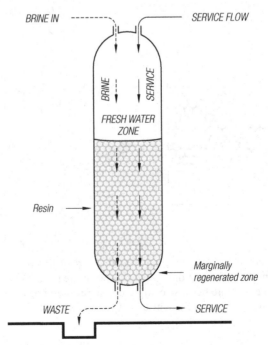

FIGURE 4-10   Cocurrent flow of regenerant brine through a typical water softener.

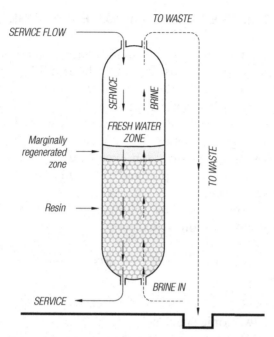

FIGURE 4-11 Countercurrent flow of regenerant brine through a typical water softener.

completely as it would be for countercurrent regeneration because these ion exchange beads have only been exposed to regenerant solution that has already reacted somewhat with beads near the top of the bed.

In the countercurrent method, the service flow moves from the top of the resin tank downward through the resin (Fig. 4-11). Regeneration is performed in the opposite direction (upflow). This method of regeneration allows the lower or bottom portion of the resin bed to receive a strong brine solution, which will initially regenerate that portion of the bed more thoroughly. When the softening unit has completed the regeneration cycle and is returned to service, the water leaving the unit passes through the most completely regenerated portion of the resin bed last. This results in a higher softening capacity with a very low hardness leakage.

For a countercurrent brining system to be effective, specific engineering conditions must be met. In particular, the bed must remain in a completely fixed position during the brining (regeneration). This allows the brine solution to be in close contact with each resin bead, thus removing the maximum number of hardness ions. In general, countercurrent regeneration will provide more efficient use of regenerant and more reliable softening, with less hardness leakage.

MATHEMATICAL TREATMENT OF ION EXCHANGE EQUILIBRIA

Readers who are not interested in mathematical analysis of chemical phenomena may skip over this section without hindering their understanding of the remainder of the text. For the mathematically inclined, this section will provide a deeper understanding of how ion exchange resins function chemically. The key concept illustrated in this section is that ion exchange processes are a well-defined, well-understood chemical phenomenon. As such, ion exchange reactions are governed by chemical equilibrium principles and can be mathematically modeled in a quantitative manner.

We begin by considering the following generalized ion exchange equilibrium, for any two ionic species:

$$|Z_B|A + |Z_A|R\text{–}B \leftrightarrow |Z_B|R\text{–}A + |Z_A|B \tag{1}$$

where:

A and B are two ionic species,

$|Z_A|$ and $|Z_B|$ are the absolute values of the charges on A and B respectively, and

R denotes the ion exchange resin polymer.

We can define the equilibrium constant $K_{A,B}^C$ for the ion exchange equilibrium:

$$K_{A,B}^C = \frac{[N_A]^{|Z_B|}[B]^{|Z_A|}}{[A]^{|Z_B|}[N_B]^{|Z_A|}} \tag{2}$$

where $[N_A]$ and $[N_B]$ represent the concentration of ions A and B, respectively, loaded onto the resin, in units of moles per liter of wetted resin. The letter "C" in $K_{A,B}^C$ denotes that we are using concentrations rather than activities, and "A,B" denotes that we are modeling the ion exchange equilibrium shown in Eq. 1 **as written**. In the following analysis we find it useful to work with equivalents, so we define:

$$\text{Equivalents of "A"} = |Z_A|[A]$$

The ionic charge is in units of equivalents/mole, so the product $|Z_A|[A]$ will have units of equivalents/liter. We can then define the total ionic strength of the solution, C_{TOT} as:

$$C_{TOT} = |Z_A|[A] + |Z_B|[B] \tag{3}$$

We can also define the equivalent ionic fraction, C_A' of species A:

$$C_A' = \frac{|Z_A| [A]}{C_{TOT}} \tag{4}$$

and similarly:

$$C_B' = \frac{|Z_B| [B]}{C_{TOT}} \tag{5}$$

The parameters C_A' and C_B' will be, by definition, unitless. We can also see that by definition:

$$C_A' + C_B' = 1 \tag{6}$$

Rearranging Eq. 4 gives:

$$[A] = \frac{C_A' \, C_{TOT}}{|Z_A|} \tag{7}$$

Similarly, rearranging Eq. 5 gives:

$$[B] = \frac{C_B' \, C_{TOT}}{|Z_B|} \tag{8}$$

Substituting Eqs. 7 and 8 into Eq. 2 gives:

$$K_{A,B}^C = \frac{[N_A]^{|Z_B|} \left(\dfrac{C_B' C_{TOT}}{|Z_B|} \right)^{|Z_A|}}{\left(\dfrac{C_A' C_{TOT}}{|Z_A|} \right)^{|Z_B|} [N_B]^{|Z_A|}} \tag{9}$$

We next consider the quantity of ions loaded onto the resin. In 1 liter of wetted resin we have N_A moles of R–A and N_B moles of R–B, so we can define the total number of equivalents, N_{TOT}, loaded onto the resin:

$$N_{TOT} = |Z_A| [N_A] + |Z_B| [N_B] \tag{10}$$

We can then define the equivalent ionic fraction of species A loaded onto the resin as N'_A

$$N'_A = \frac{|Z_A|[N_A]}{N_{TOT}} \tag{11}$$

We see that Eq. 11 is analogous to Eq. 4, with Eq. 4 describing the speciation in the aqueous phase and Eq. 11 describing the speciation of ions loaded onto the resin. We can also define the equivalent ionic fraction of species B loaded onto the resin, N'_B:

$$N'_B = \frac{|Z_B|[N_B]}{N_{TOT}} \tag{12}$$

By inspection of Eqs. 10, 11, and 12 we see that:

$$N'_A + N'_B = 1 \tag{13}$$

which is analogous to Eq. 6. Rearranging Eqs. 11 and 12 we obtain:

$$[N_A] = \frac{N'_A N_{TOT}}{|Z_A|} \tag{14}$$

and

$$[N_B] = \frac{N'_B N_{TOT}}{|Z_B|} \tag{15}$$

Substituting Eqs. 14 and 15 into Eq. 9 we obtain:

$$K^C_{A,B} = \frac{\left(\dfrac{N'_A N_{TOT}}{|Z_A|}\right)^{|Z_B|} \left(\dfrac{C'_B C_{TOT}}{|Z_B|}\right)^{|Z_A|}}{\left(\dfrac{C'_A C_{TOT}}{|Z_A|}\right)^{|Z_B|} \left(\dfrac{N'_B N_{TOT}}{|Z_B|}\right)^{|Z_A|}} \tag{16}$$

which can be simplified to:

$$K^C_{A,B} = \frac{(N'_A N_{TOT})^{|Z_B|} (C'_B C_{TOT})^{|Z_A|}}{(C'_A C_{TOT})^{|Z_B|} (N'_B N_{TOT})^{|Z_A|}} \tag{17}$$

We can simplify further if we use the relation:

$$\frac{x^a}{x^b} = x^{a-b} \tag{18}$$

so:

$$\frac{N_{TOT}^{|Z_B|}}{N_{TOT}^{|Z_A|}} = N_{TOT}^{(|Z_B|-|Z_A|)} \tag{19}$$

We can therefore rearrange Eq. 17 to:

$$K_{A,B}^C = \left(\frac{N_A'}{C_A'}\right)^{|Z_B|} \left(\frac{C_B'}{N_B'}\right)^{|Z_A|} N_{TOT}^{(|Z_B|-|Z_A|)} C_{TOT}^{(|Z_A|-|Z_B|)} \tag{20}$$

We also know that:

$$x^{-a} = \frac{1}{x^a} \tag{21}$$

so:

$$N_{TOT}^{(|Z_B|-|Z_A|)} = \frac{1}{N_{TOT}^{(|Z_A|-|Z_B|)}} \tag{22}$$

and we can use Eq. 22 to simplify Eq. 20 to yield:

$$K_{A,B}^C = \left(\frac{N_A'}{C_A'}\right)^{|Z_B|} \left(\frac{C_B'}{N_B'}\right)^{|Z_A|} \left(\frac{C_{TOT}}{N_{TOT}}\right)^{(|Z_A|-|Z_B|)} \tag{23}$$

Equation 23 is the useful result, which can be used to model any two-species ion exchange system. Moreover, if we assume plug flow through an ion exchange column, then Eq. 23 can be readily applied, using finite element analysis to model the performance of a full-scale unit.

It has long been known that ion exchange resins tend to load monovalent ions onto the resin under high ionic strength conditions and divalent (or higher valence) ions at low ionic strength. Equation 23 provides an explanation for such a tendency. If we consider a system with a given (fixed) quantity of some ion exchange resin (i.e.,

N_{TOT} = constant) in equilibrium with a solution whose ratio $C'_A : C'_B$ is fixed, then we will have:

$$K^C_{A,B} \propto \frac{N'_A{}^{|Z_B|}}{N'_B{}^{|Z_A|}} C_{TOT}^{(|Z_A|-|Z_B|)} \tag{24}$$

or, since $K^C_{A,B}$ is a constant, we have:

$$\frac{N'_B{}^{|Z_A|}}{N'_A{}^{|Z_B|}} \propto C_{TOT}^{(|Z_A|-|Z_B|)} \tag{25}$$

Thus, for example, if $|Z_A| > |Z_B|$ then an increase in C_{TOT} (i.e., the total ionic strength of the system) will result in a larger ratio:

$$\frac{N'_B{}^{|Z_A|}}{N'_A{}^{|Z_B|}}$$

which indicates that the lower valence ion (B) will be preferentially loaded onto the resin. Equation 23 thus explains why water softeners function so effectively—divalent calcium and magensium ions are loaded onto the resin when dilute hard water is passed through the softener, while the monovalent potassium or sodium ions are loaded onto the resin when the regenerant solution (which is a strong brine) is passed through the column during the regeneration cycle.

It must be noted that Eq. 23 was derived for Eq. 1 as written, that is, for the loading of species A onto the resin and stripping species B from the resin. The equilibrium constant for the reverse reaction would be the inverse of that derived, that is,

$$K^C_{A,B} = \frac{1}{K^C_{B,A}} \tag{26}$$

Values of $K^C_{A,B}$ have been determined for a wide range of resins (Table 4-1), and good data exist in the literature (Wheaton and Bauman, 1951; Bonner and Smith, 1957). Equilibrium constants for ion exchange reactions must, by their nature, be presented as couples so a frame of reference is required. Most authors use the same standard as is shown in Table 4-1, that is, equilibrium constants are given for various ions relative to lithium. The tabulated values of $K^C_{A,B}$ therefore apply to the reaction:

$$A^{n+} + nR\text{–Li} \rightleftharpoons R_n\text{–A} + Li^+ \tag{27}$$

TABLE 4-1: Equlibrium constants, $K_{A,B}^C$ for strongly acidic, polystyrene-DVB cation exchange resins. In each case, the reference ion is Li^+

Ion	4% DVB	8% DVB	16% DVB
Li^+	1.00	1.00	1.00
H^+	1.32	1.27	1.47
Na^+	1.58	1.98	2.37
NH_4^+	1.90	2.55	3.34
K^+	2.27	2.90	4.50
Rb^+	2.46	3.16	4.62
Cs^+	2.67	3.25	4.66
Ag^+	4.73	8.51	22.9
Tl^+	6.71	12.4	28.5
Mg^{2+}	2.95	3.29	3.51
Zn^{2+}	3.13	3.47	3.78
Co^{2+}	3.23	3.74	3.81
Cu^{2+}	3.29	3.85	4.46
Cd^{2+}	3.37	3.88	4.95
Ni^{2+}	3.45	3.93	4.06
Ca^{2+}	4.15	5.16	7.27
Pb^{2+}	6.56	9.91	18.0
Ba^{2+}	7.47	11.5	20.8

Equilibrium constants for all couples can then be calculated by using the relation:

$$K_{A,B}^C = \frac{K_{A,Li}^C}{K_{B,Li}^C} \qquad (28)$$

Values of equilibrium constants for anions (on anion exchange resins) can be calculated and reported in a similar manner. Equilibrium constants for anions are generally tabulated with respect to chloride (Table 4-2).

Example Calculation: A strongly acidic cation exchange resin (8% DVB) is in equilibrium with a solution consisting of 0.001798 M Ca^{2+} and 0.0020 M Na^+. Total capacity of the resin is 0.9 eq./liter. What is the loading of Ca^{2+} and Na^+ on the resin?

Solution: The equilibrium involved in this system will be:

$$Ca^{2+} + 2R\text{–}Na^+ \rightleftharpoons R_2\text{–}Ca^{2+} + 2Na^+$$

TABLE 4-2: Equlibrium constants, $K_{A,B}^C$ for strongly basic (trimethylamine) polystyrene-DVB anion exchange resins. In each case, the reference ion is Cl⁻

Ion	Equilibrium Constant
Salicylate	32.0
Iodide	8.7
Phenoxide	5.2
Nitrate	3.8
Bromide	2.8
Nitrite	1.2
Bisulfite	1.3
Cyanide	1.6
Chloride	1.0
Hydroxide	0.05–0.07
Bicarbonate	0.3
Formate	0.2
Acetate	0.2
Fluoride	0.09
Sulfate	0.15

Applying Eq. 23 we have:

$$K_{Ca,Na}^C = \left(\frac{N'_{Ca}}{C'_{Ca}}\right)^{|Z_{Na}|} \left(\frac{C'_{Na}}{N'_{Na}}\right)^{|Z_{Ca}|} \left(\frac{C_{TOT}}{N_{TOT}}\right)^{(|Z_{Ca}|-|Z_{Na}|)} \tag{29}$$

and we know that:

$$|Ca^{2+}| = 2 \text{ and } |Na^+| = 1$$

so Eq. 29 can be written:

$$K_{Ca,Na}^C = \left(\frac{N'_{Ca}}{C'_{Ca}}\right)^1 \left(\frac{C'_{Na}}{N'_{Na}}\right)^2 \left(\frac{C_{TOT}}{N_{TOT}}\right)^{(2-1)} \tag{30}$$

Next, we need to calculate values of $C'_{Ca}, C'_{Na}, C_{TOT}$, and $K_{Ca,Na}^C$ in order to be able to solve Eq. 30. From Table 4-1 we know that $K_{Ca,Li}^C = 5.16$ and $K_{Na,Li}^C = 1.98$. From Eq. 28 we can therefore calculate:

$$K_{Ca,Na}^C = \frac{K_{Ca,Li}^C}{K_{Na,Li}^C} = \frac{5.16}{1.98} = 2.606$$

We can solve for C_{TOT} using the given concentrations of the two ions present:

$$C_{TOT} = |Z_{Ca}|[Ca] + |Z_{Na}|[Na]$$

$$C_{TOT} = \left(\frac{2\,\text{eq.}}{\text{mole}}\right)\left(0.001798\,\frac{\text{mole}}{\text{liter}}\right) + \left(\frac{1\,\text{eq.}}{\text{mole}}\right)\left(0.0020\,\frac{\text{mole}}{\text{liter}}\right)$$

$$C_{TOT} = 0.005596\,\frac{\text{eq.}}{\text{liter}}$$

We can then calculate the equivalent ionic fractions for the two species present:

$$C'_{Ca} = \frac{|Z_{Ca}|[Ca^{2+}]}{C_{TOT}} = \frac{\left(2\,\frac{\text{eq.}}{\text{mole}}\right)\left(0.001798\,\frac{\text{mole}}{\text{liter}}\right)}{0.005596\,\frac{\text{eq.}}{\text{liter}}} = 0.6426$$

$$C'_{Na} = \frac{|Z_{Na}|[Na^{+}]}{C_{TOT}} = \frac{\left(1\,\frac{\text{eq.}}{\text{mole}}\right)\left(0.002000\,\frac{\text{mole}}{\text{liter}}\right)}{0.005596\,\frac{\text{eq.}}{\text{liter}}} = 0.3574$$

Substituting the values for C'_{Ca}, C'_{Na}, C_{TOT} and $K^C_{Ca,Na}$ derived above into Eq. 30 we obtain:

$$K^C_{Ca,Na} = 2.606 = \left(\frac{N'_{Ca}}{0.6426}\right)\left(\frac{0.3574}{N'_{Na}}\right)^2\left(\frac{0.005596\,\frac{\text{eq.}}{\text{liter}}}{0.90\,\frac{\text{eq.}}{\text{liter}}}\right)$$

which we can simplify:

$$2108. = \frac{N'_{Ca}}{(N'_{Na})^2} \tag{31}$$

To solve Eq. 31 we need an additional piece of information; this is provided by Eq. 13:

$$N'_{Ca} + N'_{Na} = 1 \tag{13}$$

or:

$$N'_{Ca} = 1 - N'_{Na} \qquad (32)$$

Substituting Eq. 32 into Eq. 31 gives:

$$2108. = \frac{1 - N'_{Na}}{(N'_{Na})^2}$$

or:

$$2108(N'_{Na})^2 + N'_{Na} - 1 = 0$$

Solving the quadratic we obtain:

$$N'_{Na} = 0.0215$$

and from Eq. 32 we can then calculate:

$$N'_{Ca} = 0.978$$

We can therefore conclude that at equilibrium the ion exchange resin will be 97.8% loaded with calcium, on an equivalent basis. We can also calculate the loading on a mole basis:

$$N_{Na} = \frac{N'_{Na} N_{TOT}}{|Z_{Na}|}$$

$$N_{Na} = \frac{(0.0215)\left(0.900 \dfrac{eq.}{liter}\right)}{1 \dfrac{eq.}{mole}}$$

$$N_{Na} = 0.0194 \frac{mole}{liter}$$

Similarly, we can calculate:

$$N_{Ca} = 0.440 \frac{mole}{liter}$$

and therefore, on a percentage basis:

$$\%Na = \frac{0.0194}{0.0194 + 0.440} \times 100\% = 4.22\%$$

We can therefore conclude that the resin is 4.22% loaded into the sodium form on a mole basis.

Those who have persisted in the preceding mathematical analysis will appreciate that clearly there is nothing magical about the ion exchange process. Rather, the exchange of counterions on an ion exchange resin is a simple chemical process governed by chemical equilibrium principles. These principles have shown that softening is effective because dilute solutions tend to favor adsorption of divalent ions (like calcium and magnesium in hard water) while strong solutions favor adsorption of monovalent ions (like sodium and potassium during regeneration). Finally, we have seen that the selectivity of ion exchange reactions can be modeled effectively mathematically, so we can quantify the differences in performance of sodium versus potassium regenerants; this subject is examined in the following section.

SELECTIVITY OF ION EXCHANGE REACTIONS

In the present chapter we have examined ion exchange reactions in some detail. Functional groups on the ion exchange resin attract oppositely charged ions, and these ions can be exchanged in a useful way. In the case of home softening, the ion exchange resin contains a strong acid (sulfonic acid) functional group that attracts exchangeable cations such as Ca^{2+}, Mg^{2+}, K^+, and Na^+. The chemical process of ion exchange is well understood in terms of chemical equilibrium, and it can be mathematically modeled as shown in the preceding section. Moreover, the performance of an ion exchange resin—the affinity of a resin for a particular type of ion—can be described in terms of equilibrium constants as shown in Tables 4-1 and 4-2.

The data shown in Table 4-1 are worth some further discussion. Without going into mathematical rigor, we can understand that the equlibrium constants $K_{A,B}^C$ listed in Table 4-1 are a measure of the affinity of the resin for each type of ion, relative to lithium (Li^+). For example, we see that the equilibrium constant $K_{A,B}^C$ is 3.29 for Mg^{2+} for an ion exchange resin with 8% DVB content. The equilibrium constant is greater than 1, so there is a higher affinity for Mg^{2+} than Li^+ for this

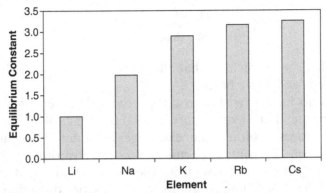

FIGURE 4-12 Equlibrium constants, $K_{A,B}^C$ for strongly acidic, polystyrene-DVB cation exchange resins with 8% DVB content for the series of alkali metal ions. The atomic mass (size) increases from left to right.

resin. The extent of the affinity for Mg^{2+} can be described mathematically with Eq. 23.

The elements lithium, sodium, potassium, rubidium, and cesium all belong to the family of alkali metals, but with progressively higher atomic weights (Chapter 5). Chemically, elements that belong to the same family behave in a similar way and have similar properties. In the present text we are interested in the differences (and similarities) of sodium and potassium chloride regenerants, so it is insightful to examine the affinity of these metal ions toward ion exchange resins. The atomic size and weight increases in the series:

$$Li^+ < Na^+ < K^+ < Rb^+ < Cs^+$$

and examination of Table 4-1 shows that as the atomic size increases, so does the affinity of the ion toward the ion exchange resin. The relationship between size (atomic number) and resin affinity is illustrated in Figure 4-12. The increasing affinity toward the ion exchange resin with increasing atomic number is noteworthy. In particular, we see that potassium has a higher affinity for the ion exchange resin than sodium, which means that K^+ will be more efficient than Na^+ at displacing hardness ions during regeneration. The better efficiency of K^+ regenerant is the underlying reason why KCl can generally be substituted for an equal weight of NaCl despite having a higher molecular weight. This point is discussed in detail in Chapters 7 and 8.

REFERENCES

Wheaton, Bauman. *Ind Eng Chem* 1951; 45: 1088.

Bonner, Smith. *J Phys Chem* 1957; 61: 326.

BIBLIOGRAPHY

Abe M, Kataoka T, Suzuki T, editors. *New Developments in Ion Exchange.* International Conference on Ion Exchange; Tokyo, Japan; 1991.

Applebaum SP. *Demineralization by Ion Exchange.* Academic Press; 1968.

Arden TV. *Water Purification by Ion Exchange.* Butterworth; 1968.

Dorfner K. *Ion Exchange Properties and Applications.* 3rd edition. Ann Arbor Science Publishers; 1972.

Faust SD, Aly OM. *Chemistry of Water Treatment.* 2nd edition. Lewis Publishers; 1998.

Helfferich F. *Ion Exchange.* McGraw-Hill; 1962.

Marinsky JA. *Ion Exchange.* Volume 1. Marcel Dekker; 1966.

CHAPTER 5

BASIC CHEMISTRY OF ION EXCHANGE

In this chapter we examine some of the underlying chemical principles required to fully understand how ion exchange processes work. Although not required for a good practical understanding of ion exchange methods, these principles will enable the reader to more fully understand and appreciate how, and why, the water softening methods work the way they do.

THE BUILDING BLOCKS OF MATTER

All matter that we see, feel, and taste in our everyday lives is composed of atoms and molecules. Atoms are the smallest possible unit of an individual element that can exist in a chemical process. Long ago, it was believed that atoms were the smallest possible units of matter, and were indivisible. This notion has long since been disproven, and it is known that atoms consist of three key building blocks: electrons, protons, and neutrons.[1]

[1]To be precise, it is well known that these three types of particles are not the ultimate, fundamental, building blocks of nature. They can each be broken into smaller subatomic particles. However, for the purposes of this text, we need not be concerned with subatomic physics, and we shall restrict our attention to the three particles discussed.

Electrons are the smallest of the three subatomic particles that we will discuss. The mass of a single electron at rest is just 9.1091×10^{-28} grams. That is, it would take 1.0978×10^{27} (1 followed by 27 zeros) electrons to have a mass of just 1 gram. Despite their small size, electrons occupy the vast majority of the space in the atom as they move around the nucleus in a complex manner. The illustrations provided in this text show the electrons moving around each nucleus in a circular orbit; such an illustration is an oversimplification of the actual electron motion, but it illustrates the general concept. In reality the electrons are delocalized over a region of space in the vicinity of the nucleus, in what is called the atomic orbital. Each electron has a single negative charge.

Protons are much more massive than electrons; each proton has 1836.1 times the mass of an electron. Despite their larger mass, protons each have only a single positive charge, equal in size to that of the electron but opposite in sign. The protons reside within the tiny nucleus of the atom, and therefore a nucleus with Z protons will have a nuclear charge of Z^+. The number of protons in the nucleus is known as the atomic number, and it determines what type of element the atom will be. For example, all hydrogen atoms have 1 proton in the nucleus; all chlorine atoms have 17 protons, etc. The total nuclear charge of Z^+ in a neutral atom will be balanced by Z electrons in the orbitals surrounding the nucleus. Electrically, the nuclear charge is balanced with the charge of the electrons, giving a balanced, neutral atom.

Neutrons also reside in the nucleus. They are very similar in mass to protons, but do not have any charge. For an atom of a given type of element the number of protons will be fixed, and equal to the nuclear charge. The number of neutrons, however, can vary. For example, all hydrogen atoms will have one proton, but different hydrogen atoms can have different numbers of neutrons. Most hydrogen atoms have no neutrons, but a small percentage of them will have one (deuterium) while an even smaller percentage will have two (tritium). Atoms that have the same atomic number (i.e., the same number of protons) but different numbers of neutrons are called different isotopes of the same element.

The mass of the proton and the neutron are very similar, and much larger than that of an electron. Therefore, it has become accepted practice to describe an atom in terms of its atomic mass, which is the sum $(Z + N)$ of the number of protons (Z) and the number of neutrons (N). Different isotopes of the same element will therefore have the same number of protons but different atomic masses. For example, all potassium atoms contain 19 protons. However, naturally occurring

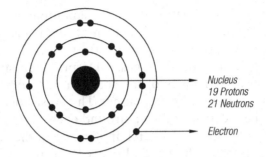

FIGURE 5-1   Illustration of an atom of the potassium isotope with atomic mass of 40.

potassium consists of three isotopes; 93.10% of the potassium atoms contain 20 neutrons and have an atomic mass of 39. Similarly, 0.0118% of the potassium ions have 21 neutrons and an atomic mass of 40, while 6.88% of them have 22 neutrons and an atomic mass of 41. Specific isotopes are typically shown with the total atomic mass superscripted and to the left of the atomic symbol. Thus the three naturally occurring isotopes of potassium are written as ^{39}K, ^{40}K and ^{41}K. A neutral potassium atom, with atomic mass of 40 is illustrated in Figure 5-1.

ATOMIC AND MOLECULAR WEIGHTS

We saw in the above section that all atoms of a given element will have the same number of protons in the nucleus and therefore the same atomic number. However, different atoms of the same element will have different atomic masses, because of the different masses of various isotopes. To standardize methods of measuring the mass of various elements, scientists adopted a convention in which the mass of the ^{12}C isotope (containing 6 protons and 6 neutrons) was defined as being 12.000 (exactly). The mass of all other atoms was then measured relative to the ^{12}C standard.

To account for the different mass of the various isotopes of an element, scientists measure the relative amounts in nature of each of the isotopes and then calculate a weighted average. For example, naturally occurring potassium consists of three isotopes:

— 93.10% of potassium ions are ^{39}K with an atomic mass of 39
— 0.0118% of potassium ions are ^{40}K with an atomic mass of 40
— 6.88% of potassium ions are ^{41}K with an atomic mass of 41

The atomic mass of potassium is therefore the weighted average of these three isotopes:

$$\text{Atomic Mass} = 0.9310 \times 39 + 0.000118 \times 40 + 0.0688 \times 41$$

$$\text{Atomic Mass} = 39.135^2$$

To do proper bookkeeping on chemical reactions, chemists have determined the number of atoms in a given mass of material. The mass of an element (measured in grams) equal to its atomic mass is said to contain one mole of this element. One mole contains 6.02214×10^{23} atoms. For example, iron has an atomic mass of 55.847. Therefore, 55.847 grams of naturally occurring iron is said to be one mole of iron, and will contain 6.02214×10^{23} iron atoms. Mathematically, we will then describe the atomic mass of iron as being 55.847 g/mole.

The concepts of the mole and atomic weights have been extended to include chemical compounds. The molecular weight for a compound is equal to the sum of all the atomic weights for each element in the compound. Each atomic weight must be multiplied by the number of times that element is found in the compound. For example, reaction of hydrogen (H, atomic mass 1.008 g/mole) with fluorine (F, atomic mass 18.998 g/mole) yields the compound HF. Each molecule of HF contains one atom of hydrogen and one atom of fluorine. Therefore the molecular weight of HF is calculated:

$$\text{Molecular Weight for HF} = 1.008 \text{ g/mole} \times 1 + 18.998 \text{ g/mole} \times 1$$

$$\text{Molecular Weight for HF} = 20.006 \text{ g/mole}$$

Similarly, one mole of carbon (C, atomic mass 12.011 g/mole) reacts with two moles of oxygen (O, atomic mass 15.999 g/mole) to form carbon dioxide, CO_2. The molecular weight of CO_2 can then be calculated:

$$\text{Molecular Weight for } CO_2 = 12.011 \text{ g/mole} \times 1 + 15.999 \text{ g/mole} \times 2$$

$$\text{Molecular Weight for } CO_2 = 44.009 \text{ g/mole}$$

Tables 5-1 and 5-2 provide summaries of atomic and molecular weights, respectively, that are often useful in water treatment applications.

---

[2]To be precise, the slightly different mass of protons and neutrons will lead to some minor differences between atomic weights calculated as shown above and the actual values. The concepts illustrated in this section, however, are correct. The precise atomic weight for potassium is 39.098.

TABLE 5-1: Atomic weights for some elements used in water treatment calculations

Element	Symbol	Atomic Weight (grams per mole)
Hydrogen	H	1.008
Lithium	Li	6.941
Boron	B	10.81
Carbon	C	12.011
Nitrogen	N	14.007
Oxygen	O	15.999
Fluorine	F	18.998
Sodium	Na	22.990
Magnesium	Mg	24.305
Aluminum	Al	26.982
Silicon	Si	28.086
Phosphorous	P	30.974
Sulfur	S	32.06
Chlorine	Cl	35.453
Potassium	K	39.098
Calcium	Ca	40.08
Iron	Fe	55.847
Copper	Cu	63.546
Zinc	Zn	65.39
Bromine	Br	79.904
Strontium	Sr	87.62
Iodine	I	126.905

TABLE 5-2: Molecular weights for some substances used in water treatment calculations

Substance	Formula	Molecular Weight (grams per mole)
Sodium chloride	$NaCl$	58.443
Sodium carbonate	Na_2CO_3	105.988
Sodium bicarbonate	$NaHCO_3$	84.006
Sodium sulfate	Na_2SO_4	142.036
Potassium chloride	KCl	74.551
Calcium oxide (lime)	CaO	56.079
Calcium hydroxide (slaked lime)	$Ca(OH)_2$	74.094
Calcium carbonate (limestone)	$CaCO_3$	100.088
Calcium magnesium carbonate (dolomite)	$CaCO_3 \cdot MgCO_3$	184.401
Calcium sulfate dihydrate (gypsum)	$CaSO_4 \cdot 2H_2O$	172.166
Calcium chloride	$CaCl_2$	110.986
Magnesium chloride	$MgCl_2$	95.211
Aluminum sulfate	$Al_2(SO_4)_3 \cdot 16H_2O$	630.372

CATIONS AND ANIONS

We saw above that an atom contains a nucleus, which consists of Z protons and N neutrons. Each proton has a single positive charge, so that the nucleus has a total charge of Z^+. In a neutral atom, there will also be Z electrons surrounding the nucleus. Each electron has a single negative charge, so the total charge of the electrons will be Z^-. Electrically, the Z^+ charge from the nucleus will be balanced by the Z^- electrical charge, leaving a neutral atom with no net electrical charge.

In many cases, atoms are not found in nature as neutral atoms. Instead, the stable form of an element may be an ion, in which one or more electrons has been added to, or removed from, the neutral atom. The sodium ion is a good example of the formation of charged ions. A neutral sodium atom consists of 11 protons and 12 neutrons in the nucleus (for the ^{23}Na isotope). The nuclear charge of +11 is neutralized by 11 electrons in orbitals surrounding the nucleus. Neutral sodium atoms can be synthesized in the laboratory, but in nature they are exceedingly rare because of the stability of the monopositive sodium ion, Na^+. Sodium ions contain the same nucleus as sodium atoms, but the orbitals around the nucleus contain only 10 electrons, which leaves the sodium ions with a net ionic charge of +1.

Many ions, like Na^+ are positively charged, and are the result of the loss of an electron from a neutral atom. Such positively charged ions are known as cations. Similarly, some types of atoms prefer to acquire excess electrons and therefore form negatively charged ions known as anions. One common example of an anion is the chloride anion. Chlorine has an atomic number of 17, and therefore a neutral chlorine atom has 17 electrons. However, chlorine has a strong tendency to acquire one additional electron, so most chlorine found in nature is in the form of the chloride anion, Cl^-, which contains 18 electrons.

Some ions are formed by the loss or acquisition of more than one electron per atom. For example, calcium and magnesium atoms tend to release two electrons to form the divalent Ca^{2+} and Mg^{2+} ions, respectively.

Ions are electrically charged, and so they will be attracted to oppositely charged species. Ion exchange resins work because the functional groups attached to the resin are charged, and so they will attract oppositely charged ions from the solution. The attraction between ions in solution and ion exchange resins does not lead to the formation of outright chemical bonds. Instead, a weak electrostatic attraction exists between the ion and the resin so that the ion can be replaced by other

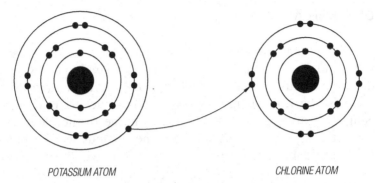

POTASSIUM ATOM   CHLORINE ATOM

FIGURE 5-2 Reaction of a potassium atom with a chlorine atom. Such a reaction will lead to the formation of the potassium chloride salt.

ions in solution; that is, the ions weakly attached to an ion exchange resin are exchangeable.

CHEMICAL REACTIONS

Chemical reactions occur when there is a transfer or a sharing of electrons between two or more atoms. For example, potassium atoms tend to be unstable, and release one electron to form the monopositive K^+ ion. Similarly, chlorine atoms tend to acquire one electron to form the Cl^- anion. A chemical reaction can then occur between a potassium atom and a chlorine atom when a single electron is transferred from the potassium atom to the chlorine atom (Fig. 5-2). In the case of reaction between potassium and chlorine the electron transfer is essentially complete (in other cases atoms tend to share electrons, rather than an outright transfer).

Transfer of an electron between a potassium atom and a chloride atom results in the formation of K^+ and Cl^- ions. There will be an electrical attraction between the positively charged potassium cations and the negatively charged chloride anions. This electrical attraction will then cause a collection of K^+ and Cl^- ions to form the salt K^+Cl^-, or potassium chloride. Salts such as potassium chloride tend to dissolve readily in water, to release the K^+ and Cl^- ions into solution.

BIBLIOGRAPHY

Atkins P, Jones L. *Chemical Principles*. W. H. Freeman; 1999.
Brown TL. *General Chemistry*. Charles E. Merrill Books; 1964.

Chang R. *Chemistry*. 6th ed. WCB McGraw-Hill; 1998.

Cracolice MS, Peters EI. *Basics of Introductory Chemistry*. Thomson Brooks/Cole; 2007.

Oxtoby DW, Nachtrieb NH, Freeman WA. *Chemistry—Science of Change*. 2nd ed. Saunders College Publishing; 1994.

Silverberg MS. *Chemistry—the Molecular Nature of Matter and Change*. 4th ed. McGraw-Hill; 2006.

Tro NJ. *Chemistry—A Molecular Approach*. Pearson Prentice-Hall; 2006.

CHAPTER 6

OPERATION AND TYPES OF WATER SOFTENERS

The design of water softeners has evolved over the years, and various types of units are commercially available. All commercial softeners rely on the same key elements: a cation exchange resin inside a pressure vessel along with a source of regenerant salt. However, there are many differences in the nature of the brine system and the regeneration cycle. In this chapter we examine and compare the various types of systems available.

HISTORICAL METHODS OF REGENERATION

Some of the earliest water softeners used a simple method of regeneration in which fine salt was simply poured on top of the ion exchange resin. Water flow then dissolved the fine salt, and a slug of the strong brine solution flowed through the resin beads, regenerating the resin in the process. Such a method was known as salt-in-head regeneration.

Later methods of regeneration included a tank for salt storage, and brine generation that was transferred to the ion exchange bed by pressure. Such systems achieved regeneration with full-strength brine.

Water Softening with Potassium Chloride: Process, Health, and Environmental Benefits, by William Wist, Jay H. Lehr, and Rod McEachern
Copyright © 2009 by John Wiley & Sons, Inc.

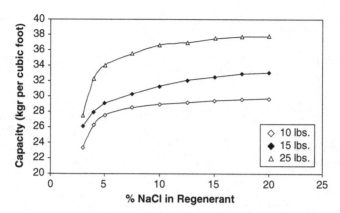

FIGURE 6-1 Softening capacity of a typical cation exchange resin as a function of the NaCl content of the regenerant salt solution for various total salt dosages.

Full-strength brine is effective as a regenerant, and contrary to popular belief it does not cause problems with cracking of the resin beads. However, regeneration with full-strength brine is not efficient use of salt. Examination of the regeneration curve (i.e., capacity of the regenerated resin) as a function of the salt concentration in the regenerant solution will typically show a diminishing returns curve for high-salt concentrations, as shown in Figure 6-1.

Figure 6-1 illustrates how the capacity of the ion exchange resin initially rises rapidly as the concentration of salt in the regenerant increases. However, above 10–15% salt, the increases in capacity are modest. As a result, 10–15% is generally deemed to be a good concentration of salt in the regenerant solution. At this concentration, good capacity of the regenerated ion exchange resin is achieved, without excessive consumption of salt.

Modern water softeners will have a vessel for storage of salt and production of the regenerant brine. In such systems, the salt brine is proportioned with water so that the regenerant has the desired 10–15% concentration of salt. In the next section, we take a closer look at a typical brine regeneration cycle.

OPERATION OF A TYPICAL WATER SOFTENER

The cycle of water and regenerant flow through an ion exchange bed in a softener is not as simple as one would initially think. In reality, there are several stages to the process, as shown in Figure 6-2. We shall examine each of the stages in some detail in the following pages. It must

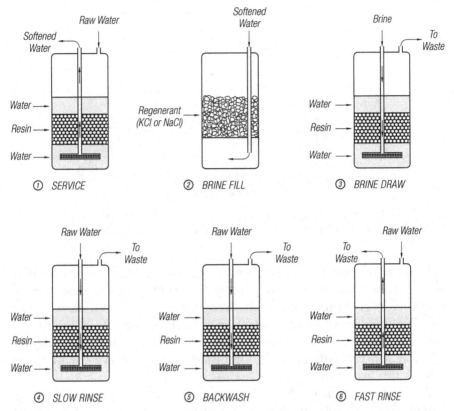

FIGURE 6-2 Typical sequence of operations for the water softening cycle. Variations in this typical sequence are available from different manufacturers, as described in the text.

be noted that there are many variations in the softening sequence, in particular with respect to the direction of the various flows (countercurrent vs. cocurrent). In this section the most common sequence for softening is discussed; differences between this and other sequences are examined in the following section.

We begin our examination of the operation with the *service cycle* (Fig. 6-3). In the service cycle the flow of water is always the same, namely, from the top down. As the water flows through the bed of ion exchange resin, it is softened and filtered. The purified water is collected from near the bottom of the ion exchange tank in a header and returned by an internal pipe to the soft water exit, usually at the top of the tank.

The next stage in the softening cycle is the *brine fill* (Fig. 6-4). In this sequence, water is added to the salt storage vessel. Salt dissolves,

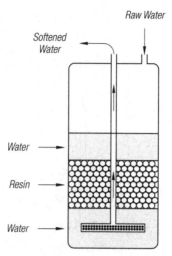

FIGURE 6-3 The service cycle. Normal operation of the unit, in which water is softened, filtered, and refined as it flows down through the resin. All softeners use the same service cycle.

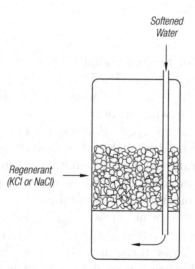

FIGURE 6-4 The brine fill, which can be at the beginning or the end of the regeneration process. During the brine fill stage, the exact amount of soft water needed is pumped into the salt tank, to create brine. The quantity of soft water can be controlled by a flowmeter or a float control valve.

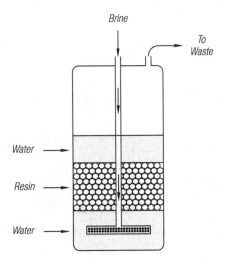

FIGURE 6-5 The brine draw stage of softening. Typically a venturi system is used to draw brine from the brine tank and mix it with water. The diluted brine is drawn from the salt tank down through the riser, flows upward through the stratified resin bed, and then reports to the drain. The brine flow removes hardness from the tank by conversion of the resin from the calcium/magnesium state back into the sodium state.

to form a saturated or near-saturated brine solution. In some systems, the exact amount of water added is controlled by a flowmeter or a float control valve. In some systems, the brine fill occurs just before the regeneration cycle, while in other cases the brine fill occurs at the end of the regeneration cycle. Performing the brine fill after the regeneration cycle would offer the advantage of more reliable control of the salt concentration during regeneration, since the brine would be consistently saturated by the time of the next regeneration cycle.

The next stage in the water softening cycle is the *brine draw* (Fig. 6-5). During the brine draw stage, brine is drawn from the brine tank by a venturi system. Water is the motive force in a venturi system; the water flows through a specially designed nozzle and aspirates (draws) the brine into the water stream. As a result, the saturated brine in the brine tank is transferred to the ion exchange tank, and diluted simultaneously to the desired concentration of 10–15% NaCl.

The diluted brine flows into the ion exchange resin tank through a riser pipe and is introduced into the tank through a brine header (Fig. 6-5). The brine flows upward through the stratified resin bed and then flows out to waste. The flow of brine regenerates the ion exchange

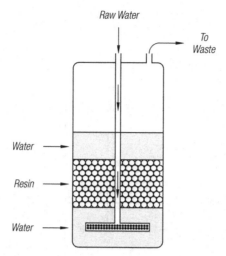

Raw Water

To Waste

Water

Resin

Water

FIGURE 6-6   The slow rinse stage. The slow rinse moves the brine through the resin bed, providing additional hardness removal. It also ensures that the brine has passed completely through the resin. The direction of flow is the same as the brine draw.

resin, from the calcium and magnesium form back into the sodium form. As a result, the waste brine will contain a mixture of $CaCl_2$ and $MgCl_2$ along with excess NaCl. As the resin is regenerated, the composition of the waste brine will change; initially it will be very rich in $CaCl_2$ and $MgCl_2$, but as regeneration proceeds toward completion, the concentration of calcium and magnesium will become lower and the concentration of excess NaCl will rise.

When the brine draw sequence is complete, the system enters the *slow rinse* stage (Fig. 6-6). At this time, raw (hard) water enters and is introduced into the resin bed in the same direction as during the brine draw. The brine in the resin bed at the end of the brine draw sequence will be a mixture of $CaCl_2$, $MgCl_2$, and NaCl, which would be undesirable in the final water product. The slow rinse ensures that the brine has passed completely through the resin, by slowly displacing and diluting the brine. As the brine displacement occurs, some further regeneration will occur, since the brine in the resin bed at the end of the brine draw sequence will be rich in excess NaCl.

After the slow rinse, the softener enters the *backwash* stage (Fig. 6-7). In the backwash stage, raw water enters the pressure vessel via the riser pipe and is introduced to the resin bed via the header, as was done in the slow rinse stage. However, in the backwash stage the flow of water is higher, so that the water flow fluidizes the bed of ion

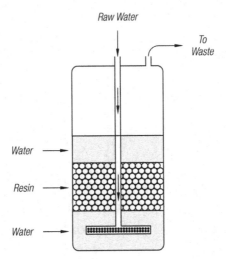

FIGURE 6-7 The backwash stage. During this stage of the softening cycle, a final countercurrent rinse cleans and stratifies the resin beads. It removes dirt and loosens and fluffs the resin. If the backwash is done after the brine draw, it helps displace any remaining regeneration brine.

exchange resin. At these high flow rates, the water is able to remove dirt and other solid-phase impurities that were filtered out of the water during the service cycle. Precipitated iron oxides will also be removed from the system during the backwash stage. Fluidizing the resin also loosens and fluffs the bed so that all of the ion exchange beads will come into contact with the flow of water. In addition to the removal of dirt and fluffing the resin, the flow of water in the backwash stage further lowers the amount of regeneration brine ($CaCl_2$, $MgCl_2$, and NaCl) in the bed.

In order for the backwash stage to be effective in fluidizing the bed of resin, the flow of water during the backwash stage must be upwards, that is, countercurrent to the flow of water during the service cycle.

The final stage of the regeneration process is the *fast rinse* (Fig. 6-8). During the fast rinse, raw water enters the pressure vessel through the normal entrance (that is, the same as during the service cycle) and exits via the riser pipe. The direction of water flow during the fast rinse is therefore the same as during the service cycle. During the fast rinse the water flow repacks the resin bed and removes any regeneration brine remaining in the system. After the fast rinse the regeneration sequence is complete, and the ion exchange system is ready to be returned to service.

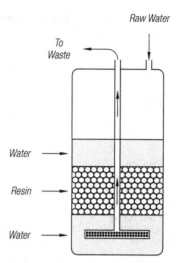

FIGURE 6-8 The fast rinse stage. This stage of the sequence repacks the resin bed and prepares the system to return to service. The direction of flow will always be the same as the service cycle. The fast rinse removes any remaining regeneration brine and leaves the unit ready for service.

COMMON SEQUENCES OF CYCLES

The description of the softening process in the above section is very common, and typical of the sequence of cycles. However, there are several variations on this sequence, as shown in Figure 6-9.

Different cycles are provided by various manufacturers of water softening systems. All such commercially available systems will function effectively, but the choice of optimal system will depend, in part, on the composition of the water to be softened. Some cycles will also provide more efficient use of regenerant, that is, gallons of water softened per pound of regenerant.

TYPES OF WATER SOFTENERS

Water softeners have evolved over time, so that there is a range of sophistication in the units in service. Historically, the tendency has been for greater automation and reliability in the units. More recently there has been a shift toward more efficient use of water and salt in the regeneration cycle to minimize the use of salt, for convenience and cost, and to reduce the impact of sodium chloride on the environment.

The earliest water softeners were generally *manual*. In a manual system, all stages of the softening sequence are controlled by manual

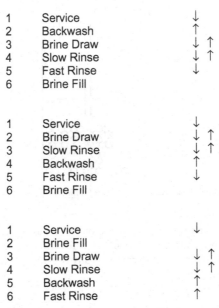

Common Sequences of Softener Cycles

FIGURE 6-9   Common sequences of the water softening cycle, available in different units provided by various manufacturers. In all cases the service cycle is downward ($\downarrow$), meaning that the water flow is down through the resin beads and then upward through the riser pipe. Flow of the brine and rinse water can be the same as during the service cycle (cocurrent, $\downarrow$) or in the opposite direction (countercurrent, $\uparrow$).

intervention. That is, the backwash, brining, rinsing, and return to service are all controlled manually by the owner turning valves in the appropriate sequence. In contrast, in *semiautomatic* softeners, the owner must only initiate the backwash and the brine cycle manually. A timer then completes the brine and rinse operations and returns the unit to service automatically.

The degree of automation is expanded in an *automatic* system. With an automatic system, the owner needs only to initiate the regeneration cycle when the quality of the softened water begins to deteriorate. All other stages of the water softening sequence are controlled by a timer. *Fully automatic* systems are also available, in which the owner does not need to initiate the regeneration cycle. Instead, a timer automatically initiates the regeneration process at preselected intervals. The length of the time interval between regeneration cycles will be a function of the capacity of the ion exchange bed and the hardness of the water to be softened.

The evolution of water softeners continued beyond fully automatic units with the introduction of *demand-initiated regeneration* (DIR) water softeners. In these units, the water softener is equipped with a flowmeter, which automatically initiates the regeneration process after a predetermined volume of water has been softened. Alternately, DIR units can be based on hardness sensors, which monitor the quality of the softened water and initiate the regeneration cycle when the quality of softened water begins to deteriorate. Demand-initiated regeneration units provide the convenience of fully automatic softeners while also conserving salt by only initiating a regeneration sequence when it is required to maintain water quality.

Modern softeners continue to improve. Most recently, computer control of the regeneration sequence has been introduced. In such models, the computer calculates the most efficient quantity of salt to be used for regeneration based on the volume of water treated, the hardness level in the water, and historical data on regeneration efficiency. The computer then controls the regeneration sequence accordingly.

All the water softening systems described above are variations of the basic system described in Chapter 4. In all cases the basic system consists of a pressure vessel containing ion exchange resin beads, along with a brine tank in which salt is stored and the brine is made. An entirely different approach can be used with *portable exchange tanks* (PET). PET systems do not have a separate tank for storage of salt and brine generation. Instead, a PET system consists of a tank or cartridge containing ion exchange resin. This tank is equipped with special fittings designed for easy connection and disconnection of the tank. The PET softeners do not include the valving or controls required for regeneration. Instead, when regeneration is required, the tank is disconnected and transported to a central station or plant for regeneration. When regeneration is complete the unit is returned, reconnected, and returned to service.

The early evolution of water softeners was toward a higher degree of automation and sophistication. More recently, the need to conserve water has lead to the development of units that use less water during the regeneration cycle. Such units incorporate features such as:

- Smaller-sized ion exchange beads.
- "Packed beds" of ion exchange resin, with no freeboard in the tank.
- Smaller tank diameters.
- The use of less ion exchange resin.
- Countercurrent regeneration.

Water softeners using these conservation features, along with demand-initiated regeneration, can reduce water consumption during the regeneration cycle between 30% and 75%. Reduction in water consumption is achieved in part by elimination of the large volume of water in the freeboard, which consumes a significant quantity of water because the brine from regeneration must be diluted and displaced before the unit can be returned to service. Water softeners with these features can be very efficient, with as little as 40 gallons of water required to regenerate a tank containing 2 cubic feet of resin. In addition to the reduction of water consumption, the more efficient units will provide greater salt efficiencies.

One issue with high-efficiency softeners that needs to be addressed is the need for upstream filtration before softening. The packed beds, with fine bead size and the elimination of the backwash stage, means that such softeners are vulnerable to plugging with dirt or precipitates. Prefiltration is required, even if the levels of suspended solids are very low, as found in the water supply of most cities.

SIZING A WATER SOFTENER

With an understanding of key concepts such as water hardness and principles of operation, it is possible to size a water softener for any application. Detailed information on sizing is available from equipment suppliers, and several good guides are available on the Internet. As an example, we will work through the sizing calculation for a typical family of four, with influent hardness of 34 grains per gallon (gpg) and no dissolved iron or manganese.

The first task is to estimate the usage of water in gallons. The water usage can be determined by measurement or estimated in detail by tabulating the number of water-using fixtures. However, many suppliers use a simple calculation in which the daily usage is estimated by assuming a fixed consumption per occupant per day. Consumption of 60 gallons per day per occupant is a typical multiplier. In addition, the number of occupants is typically increased by one to account for appliances. In the present example, the daily water consumption will be:

$$\text{Water Consumption} = (\text{number of occupants}) \times 60\,\text{gallons/day}$$

$$\text{Water Consumption} = (4+1) \times 60 = 300\,\text{gallons/day}$$

With information on the water consumption and the influent hardness, we can then calculate the total grains of hardness that needs to be removed during softening. In the present example, the influent hardness is 34 gpg with no iron or manganese, so the total daily demand for softening will be:

Daily Demand = Water Consumption × Influent Hardness

Daily Demand = 300 gallons/day × 34 gpg = 10,200 grains/day

The daily demand for softening can be used to calculate the required softener capacity. The required softener capacity will depend on the number of days between regeneration cycles. The days between cycles will be recommended by the supplier, but will typically be in the range of 2–7 days. In our sample calculation, we will use a typical value of 4 days between regeneration cycles. The softener capacity can then be calculated by multiplying the daily softening demand by the number of days between regeneration cycles:

Softener Capacity = Daily Demand × Days between Cycles

Softener Capacity = 10,200 grains/day × 4 days = 40,800 grains

The water softener can then be sized appropriately. Typically, the calculated capacity will be rounded up (to be safe), so in the present example one would recommend a softener with capacity of, for example, 45,000 grains.

In the above example, there was no iron or manganese in the influent water. If these elements are present in solution, then the influent hardness should be corrected to account for these elements, since they will adsorb onto the ion exchange resin in the same way as calcium and magnesium. The corrected hardness can be calculated by multiplying the iron content (in mg/L or ppm) by 4 to convert to grains per gallon hardness equivalent. Similarly, the manganese content can be converted from mg/L (ppm) to gpg equivalent by multiplication by 2.

Consider a second example for the same family of four. In this example, the influent hardness is again 34 gpg, but with 2.1 ppm dissolved iron and 0.5 ppm manganese. The corrected influent hardness can then be calculated:

Hardness = Hardness (gpg) + Iron (ppm) × 4 + Manganese (ppm) × 2

Hardness = 34 + (2.1 × 4) + (0.5 × 2) = 43.4 gpg

The total daily demand for softening will then be:

Daily Demand = 300 gallons/day × 43.4 gpg = 13,020 grains/day

Most suppliers recommend no more than 2 days between regeneration cycles if the influent water has an iron content of 2 ppm iron or higher. Therefore, in this example we calculate the softener capacity based on 2 days between regeneration cycles. The softener capacity would then be:

Softener Capacity = 13,020 grains/day × 2 days = 26,040 grains

As with the first example, one would generally size the water softener with some excess capacity to ensure that good softening was achieved even during very high-usage periods. In the present case, one would probably round up and size a softener with a capacity on the order of 30,000 grains.

In practice, proper sizing of a water softener is more complicated than just determining capacity requirements. The required flow rates through a softener will have an impact on selection of the volume of resin, which will in turn impact the required amount of salt (and salt efficiency). We defer this discussion until Chapter 8, where the concept of the efficient use of salt is introduced and discussed in detail.

CHAPTER 7

POTASSIUM CHLORIDE REGENERANT FOR WATER SOFTENING

Sodium chloride has been the most common regenerant used in water softening for many years. There are many good reasons why sodium chloride has become so widely used. It is mass produced all over the world, and therefore it is readily available at low cost. It is safe to work with and handle. Finally, the high solubility of sodium chloride makes it suitable as an effective regenerant for removal of calcium and magnesium ions from an ion exchange resin.

In recent years, concerns have been expressed by medical, environmental, and regulatory professionals about possible negative impacts of widespread sodium chloride usage in water softeners on human health and the environment. As a result, significant efforts have been expended on the search for alternatives to sodium chloride for water softening.

ALTERNATE REGENERANTS

Several common materials have been evaluated for use as regenerants. Some of these materials include sodium carbonate (soda ash), sodium bicarbonate (baking soda), and sodium sulfate. These materials are

Water Softening with Potassium Chloride: Process, Health, and Environmental Benefits,
by William Wist, Jay H. Lehr, and Rod McEachern
Copyright © 2009 by John Wiley & Sons, Inc.

TABLE 7-1: Cold water solubility (in grams of solute per 100 cc of water) of various salts considered for use as a water softening regenerant

Regenerant	Solubility (grams per 100 cc)
Sodium chloride (NaCl)	35.7
Sodium sulfate (Na$_2$SO$_4$)	4.76
Sodium carbonate (Na$_2$CO$_3$)	7.1
Sodium bicarbonate (NaHCO$_3$)	6.9
Potassium chloride	34.7

Data are from the CRC Handbook of Chemistry and Physics (Weast, 1977).

readily available at low cost, but their solubility is significantly lower than sodium chloride (Table 7-1), so they cannot provide the necessary driving force for effective regeneration of ion exchange resins.

Sodium citrate (Na$_3$C$_6$H$_3$O$_7 \cdot$2H$_2$O) has also been evaluated for use as a regenerant. This material has a high water solubility (72 grams per 100 cc), so it has the potential to be an effective regenerant. In addition, sodium citrate is safe to use. However, the cost of sodium citrate is relatively high (approximately $2.00 per pound), so it has not become widely used in water softening.

Sodium nitrate has been used on occasion as a water softening salt. This material has been mined extensively for well over a century in the Atacama Desert region of Chile and has been used for fertilizer, explosives, and ceramics. At one time, sodium nitrate was used locally for water softening since it was readily available and cheap (McGowan, 1988). However, concern over the health effects of nitrates, as well as their impact on the environment, has halted this practice.

Sodium hydroxide (caustic soda, NaOH) has the high solubility (42 grams per 100 cc in cold water) required for use as a regenerant. However, sodium hydroxide is relatively expensive, and solutions of NaOH are extremely corrosive, so this material has not become widely used as a regenerant.

Seawater is used as a water softening regenerant in some specialized applications, for example, in isolated resorts where seawater is abundant but freshwater is scarce. Seawater can be used as the regenerant, as well as for backflushing, thus conserving the use of freshwater. The abundance and low cost of seawater make its use attractive; however, the low salt concentration and the presence of other ions, including calcium and magnesium (see Table 7-2) pose some problems. Water softened with seawater regenerant will not be as soft as when softened

TABLE 7-2: Common elements present in seawater

Element	Symbol	Concentration (ppm)
Chlorine	Cl	18,980
Sodium	Na	10,561
Magnesium	Mg	1,272
Sulfur	S	884
Calcium	Ca	400
Potassium	K	380
Bromine	Br	65
Carbon (inorganic)	C	28
Strontium	Sr	13
Boron	B	4.6
Fluorine	F	1.4

Data are from the CRC Handbook of Chemistry and Physics (Weast, 1977).

with NaCl. In addition, the presence of hardness ions in the seawater causes the ion exchange resin to have a lower capacity. For example, if an ion exchange system uses 40 gallons of seawater to regenerate each cubic foot of ion exchange resin, then an influent hardness of 500 ppm (total cations) would yield an operating capacity of just 15,000 grains per cubic foot (McGowan, 1988).

In addition to the issue of low capacity, seawater must be adequately cleaned before use as a regenerant. In most cases filtration and chlorination would be required to avoid fouling of the ion exchange resin.

POTASSIUM CHLORIDE REGENERANT

The final alternative regenerant to be discussed is, of course, potassium chloride (KCl). As an alternative to sodium chloride, KCl is capable of being an excellent regenerant. Potassium chloride occurs naturally in large deposits throughout Canada, the United States, and elsewhere. It readily dissolves, and a saturated solution has strength (concentration) similar to that of NaCl.

However, there have been concerns that, because of the higher molecular weight of KCl (molecular weight 74.55 g/mole)[3] relative to NaCl (58.43 g/mole), KCl would be less effective as a regenerant on a per-pound basis. The ratio of molecular weights of KCl:NaCl is 1.28:1,

[3]For an explanation of atomic and molecular weights, please see Chapter 5.

which suggests that there would be a much lower capacity per pound of regenerant for KCl. Assuming that KCl and NaCl were equally efficient in removing hardness ions from a cation exchange resin, converting the regeneration compound from NaCl to KCl would result in a 28 percent increase in the required weight of the potassium chloride regenerant:

$$\left(\frac{74.55 - 58.43}{58.43}\right) \times 100\% = 27.6\%$$

Potassium chloride has been commercially available as a regeneration compound for some years. The field results when KCl was substituted pound for pound for NaCl seemed not to support this molecular weight comparison theory, since somewhat comparable capacities were obtained. Research was therefore performed to directly measure and compare the relative efficiency and capacity of potassium chloride (KCl) to sodium chloride (NaCl) when used in regenerating a water softener. The results of such research are discussed in the remainder of this chapter, and presented in detail in Chapter 8.

INITIAL COMPARISON OF KCl AND NaCl

A series of initial tests were performed to gain insight into the reported difference between KCl capacity as a regenerant versus that expected based on molecular weights. The results presented in this section introduce the reader to the key concepts, performance parameters, and results. A more detailed comparison of the performance of KCl and NaCl is provided in Chapter 8.

Apparatus

A commercially available water softener was used for the evaluation. The water softener was loaded with exactly 53 pounds of resin, which is exactly 1 cubic foot, according to the manufacturer's specifications.

The softener used for the testing featured countercurrent regeneration, or upflow of brine in the regeneration cycle and downflow of water in the service cycle.

The softener was connected to both the local domestic (City of Saskatoon) water supply as well as an 8000 US gallon tank. The test apparatus (Fig. 7-1) consisted of an inlet valve for both water supplies, water filter and pressure gauge, influent sampling port, totalizing water

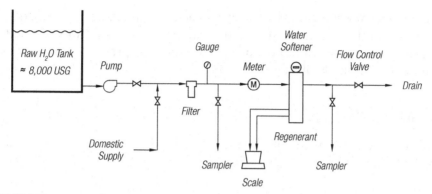

FIGURE 7-1 Apparatus for the initial study comparing the effectiveness of KCl and NaCl as softener regenerants.

meter registering flow followed by the water softener, effluent sampling port, and effluent water regulating valve.

Test Water and Regenerants

Two raw waters were used in the testing. Raw water #1 contained 20 grains/USG (as $CaCO_3$) of calcium hardness and 10 grains/USG (as $CaCO_3$) of magnesium hardness. Raw water #2 contained 20 grains/ USG (as $CaCO_3$) of magnesium hardness and 10 grains/USG (as $CaCO_3$) of calcium hardness. Calcium and magnesium hardness was added as necessary to both raw waters by addition of calcium chloride and magnesium chloride. The sodium and potassium levels in both raw waters were kept the same at 50 mg/L.

The regenerants were made up as 24% solutions, which is just slightly under the saturation point, and therefore eliminates the temperature effect on saturated brines (i.e., precipitation/dissolution of solids when temperatures fluctuate). The regenerant solutions were made up with local raw water and commercially available water softening salt for NaCl, and PotashCorp Cory Division WSM 0.2 chicklets (>99.8% KCl) for KCl.

Saskatoon city water was used for the regeneration cycle since the pump (Figure 7-1) did not have a high enough pressure to regenerate properly. When more than 8 pounds of salt was used in the regeneration cycle, the softener timer was turned off at the 8 lb level and the remaining salt eluted before the timer was turned back on to complete the regeneration cycle. This allowed the salt to pass completely through the resin bed. To get the exact weight of regenerant, a container with the regenerant and eluant tubes was placed on a balance and tared to zero. When the negative value of the required regenerant was reached, the tubes were pulled out of the regenerant. This procedure was accurate to 0.03 pounds of salt.

Analytical Methods

All water samples were analyzed for Na, K, Mg, and Ca in accordance with standard methodology (Clescerl et al., 1999).

Test Runs

Each test run was performed at a flow rate of 4 USG per minute. A typical run would consist of the following sequence:

1. Record the initial water meter reading.
2. Regenerate as described in the "Test Water and Regenerants" section.
3. Record final water meter reading of regeneration cycle, which is also the starting meter reading of the service cycle.
4. Switch valves so raw water is drawn from the tank.
5. Open flow control valve and establish a flowrate of 4.0 USG per minute.
6. Collect and analyze effluent samples for hardness by EDTA titration to accurately determine the endpoint of 1.0 grains/USG.
7. Record the final water meter reading.
8. Calculate capacity in kilograins and kilograins/lb/ft^3.

The complete sequence of 23 test runs is shown in Table 7-3.

Examination of Table 7-3 shows that the tests with 4 pounds of regenerant were repeated. The reason for repeating the 4 lb. test was to achieve stability after the heavy salt dosage in the preceding run. The results showed that the duplicate tests were almost identical, so the repeat test was used in the calculation of results.

Test Results

The total volume of hard water treated before the effluent reached a hardness of 1.0 grains/USG was measured for each test run. From this volume, and the known hardness content of the influent, the capacity

TABLE 7-3: Sequence of test runs for the initial set of tests designed to compare the performance of KCl and NaCl as softener regenerants

Run #	Regenerant	Pounds of Regenerant	Raw Water #
1	NaCl	35	1
2	NaCl	4	1
3	NaCl	4	1
4	NaCl	8	1
5	NaCl	12	1
6	NaCl	15	1
7	KCl	35	1
8	KCl	4	1
9	KCl	4	1
10	KCl	8	1
11	KCl	12	1
12	KCl	15	1
13	KCl	4	2
14	KCl	4	2
15	KCl	8	2
16	KCl	12	2
17	KCl	15	2
18	NaCl	35	2
19	NaCl	4	2
20	NaCl	4	2
21	NaCl	8	2
22	NaCl	12	2
23	NaCl	15	2

of the resin was calculated. In addition to the total resin capacity, the capacity per pound of regenerant was also calculated. The results for each of the test runs are shown in Table 7-4.

The data provided in Table 7-4 show that there was no substantial difference in softening performance between raw water #1 and #2. Therefore, the capacity data in Table 7-4 was averaged for the two raw waters; the average results are shown in Table 7-5.

The averaged capacity data shown in Table 7-5 are illustrated in Figures 7-2 and 7-3. The capacity data shown in Figures 7-2 and 7-3 suggest very little difference in performance between KCl and NaCl as regenerants on a weight basis. This conclusion is contrary to what one would expect from the ratio of molecular weights (see the section "Potassium Chloride Regenerant"). The comparison between the capacity of KCl and NaCl regenerants is explored in more detail in Chapter 8, in which we present the results from a standardized test protocol performed independently by four separate laboratories.

TABLE 7-4: Measured capacity, and capacity per pound of resin, for each test run for the initial set of tests designed to compare the performance of KCl and NaCl as softener regenerants

Run #	Regenerant	Pounds of Regenerant	Capacity (Kgrain/ft^3)	Capacity (Kgrain/ft^3/lb)
1	NaCl	35	30.8	0.88
2	NaCl	4	16.3	3.99
3	NaCl	4	16.0	3.99
4	NaCl	8	22.2	2.78
5	NaCl	12	25.6	2.13
6	NaCl	15	27.2	1.81
7	KCl	35	30.6	0.87
8	KCl	4	16.6	4.16
9	KCl	4	16.4	4.11
10	KCl	8	23.9	2.99
11	KCl	12	26.4	2.20
12	KCl	15	28.3	1.88
13	KCl	4	17.2	4.29
14	KCl	4	16.5	4.13
15	KCl	8	24.3	3.04
16	KCl	12	27.3	2.28
17	KCl	15	29.8	1.99
18	NaCl	35	31.2	0.89
19	NaCl	4	17.8	4.44
20	NaCl	4	16.4	4.10
21	NaCl	8	23.5	2.93
22	NaCl	12	27.1	2.26
23	NaCl	15	29.6	1.98

TABLE 7-5: Average capacity, and capacity per pound of resin, for each regenerant dosage and type for the initial set of tests designed to compare the performance of KCl and NaCl as softener regenerants

Regenerant	Pounds of Regenerant	Average Capacity (Kgrain/ft^3)	Average Capacity (Kgrain/ft^3/lb)
NaCl	4	16.63	4.16
NaCl	8	22.85	2.86
NaCl	12	26.35	2.20
NaCl	15	28.40	1.89
NaCl	35	31.00	0.89
KCl	4	16.68	4.17
KCl	8	24.10	3.01
KCl	12	26.85	2.24
KCl	15	29.05	1.94
KCl	35	30.60	0.87

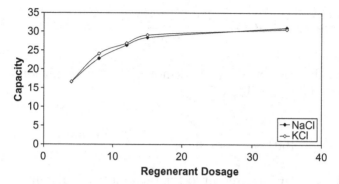

FIGURE 7-2   Comparison of the total exchange capacity (Kgrains/ft^3) as a function of regenerant dosage (pounds), comparing the performance of KCl and NaCl regenerants.

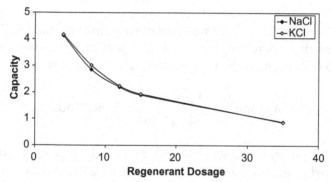

FIGURE 7-3   Comparison of the total exchange capacity per pound of regenerant (Kgrains/ft^3/pound) as a function of regenerant dosage (pounds), comparing the performance of KCl and NaCl regenerants.

It is interesting to examine the reason why KCl has a similar capacity to NaCl despite having a much higher molecular weight. We know (Chapter 3) that the softening reaction (with NaCl regenerant) is:

$$2R-SO_3^-Na^+ + Ca^{2+} \rightarrow (R-SO_3^-)_2\, Ca^{2+} + 2Na^+$$

with the regeneration reaction being the reverse. The adsorption of one ion of calcium onto the resin is associated with the displacement of two ions of sodium. Therefore we can calculate the pounds of sodium chloride required for the adsorption of one grain of calcium carbonate hardness. First, we convert from grains to grams:

$$\text{one grain } CaCO_3 \times \frac{0.064799 \text{ grams}}{\text{grain}} = 0.064799 \text{ grams}$$

Then we calculate the number of moles of calcium carbonate in this mass, using data from Table 5-2:

$$\frac{0.064799 \text{ grams}}{100.088 \text{ grams/mole}} = 0.00064742 \text{ moles}$$

Two moles of NaCl are required for each mole of calcium. Using that fact, along with the molecular weight of NaCl from Table 5-2, we can calculate the mass of NaCl required:

$$0.00064742 \text{ moles} \times 2 \times 58.443 \frac{\text{grams}}{\text{mole}} = 0.075674 \text{ grams}$$

We therefore conclude that one grain of calcium carbonate will require 0.075674 grams of NaCl regenerant. One kilograin of hardness will then require 75.674 grams of NaCl. Converting to pounds we get:

$$75.674 \text{ grams} \times \frac{1 \text{ pound}}{453.54 \text{ grams}} = 0.16685 \text{ pounds}$$

With the conversion between NaCl and kilograins of hardness as calcium carbonate in hand, we can determine what fraction of the regenerant salt was actually used in the softening process. Referring to Table 7-5, we see that for NaCl, regeneration with 4 pounds of salt gave a capacity of 16.63 Kgrain/ft³. The portion of the 4 pounds regenerant actually used in the softening process was then:

$$16.63 \text{ Kgrain} \times \frac{0.16685 \text{ pounds NaCl}}{\text{Kgrain calcium carbonate}} = 2.775 \text{ pounds NaCl}$$

The percentage of the NaCl actually used in the softening process was therefore:

$$\frac{2.775 \text{ pounds NaCl}}{4.00 \text{ pounds NaCl}} \times 100\% = 69.4\%$$

and the percentage unused NaCl will therefore be:

$$100 - 69.4 = 30.6\%$$

TABLE 7-6: Percentage of regenerant used (for softening) and unused, for the initial set of tests designed to compare the performance of KCl and NaCl as softener regenerants

Regenerant	Pounds of Regenerant	Percentage Used	Percentage Unused
NaCl	4	69.4	30.6
NaCl	8	47.7	52.3
NaCl	12	36.6	63.4
NaCl	15	31.6	68.4
NaCl	35	14.8	85.2
KCl	4	88.8	11.2
KCl	8	64.1	35.9
KCl	12	47.6	52.4
KCl	15	41.2	58.8
KCl	35	18.6	81.4

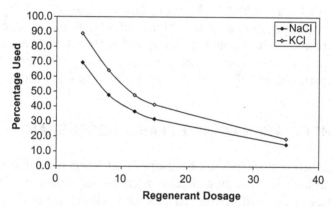

FIGURE 7-4 Percentage of regenerant used (for softening) as a function of regenerant dosage (in pounds) for KCl and NaCl.

In a similar manner, we can calculate that one kilograin of calcium carbonate would be equivalent to 3.539 pounds of KCl. We can then calculate the percentage of regenerant used and unused, for each of the regenerant types and dosages presented in Table 7-5. The results of this calculation are shown in Table 7-6.

The data in Table 7-6 are illustrated in Figures 7-4 and 7-5 for the percentage used and unused, respectively. Figures 7-4 and 7-5 show clearly why the performance of KCl is very similar to that of NaCl on a weight basis. The effectiveness of KCl as regenerant does not follow the proportions as suggested by the molecular weights (28% less effective) because the percentage of regenerant not used in softening is greater for NaCl than KCl. The KCl and NaCl regenerants have similar

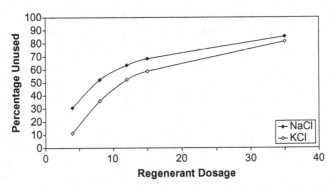

FIGURE 7-5   Percentage of regenerant unused (for softening) as a function of regenerant dosage (in pounds) for KCl and NaCl.

capacities because the percentage of KCl used for softening is significantly higher than for NaCl. The more efficient use of KCl regenerant is due to a the fact that K^+ has a higher affinity than Na^+ for the ion exchange resin, as shown in Table 4-1, and is therefore more efficient at stripping Ca^{2+} and Mg^{2+} from the resin during regeneration. This point is illustrated and discussed in more detail in Chapter 8.

CHALLENGES WHEN USING KCl AS A REGENERANT

One of the problems reported by users of KCl regenerant is bridging, or recrystallization. Bridging is also a common problem with NaCl regenerant, but the underlying reasons for bridging are quite different with KCl.

In a softener that uses NaCl regenerant, bridging occurs when there is caking of the salt above the brine level. The caking causes a void space between the brine and the salt, and undersaturated regenerant brine is produced as a consequence. In such a case inadequate softening will result. Aside from the hard water being produced, the problem can be difficult to identify because the salt compartment appears to be normal when inspected from the top—with an adequate amount of free-flowing NaCl.

Bridging can also occur with KCl regenerant, but in an entirely different way. In a softener using KCl regenerant, bridging is caused by recrystallization (not caking) and is manifested as a layer of recrystallized KCl on the bottom of the brine tank. The recrystallization can continue until a layer several inches thick forms on the bottom of the tank.

Unlike NaCl, the solubility of KCl is strongly dependent on temperature. The solubility of KCl in water increases with temperature, as shown in Figure 1-9. Recrystallization is the result of fluctuating temperatures. As the temperature of the brine increases, more KCl is dissolved. When the temperature decreases, the extra KCl that was dissolved precipitates and drops to the bottom of the brine tank. The precipitate, along with the saturated brine, forms a very hard crust, which continues to grow as the temperature fluctuates from daytime highs to nighttime lows. KCl bridging is primarily a problem when the softener is outside or in a garage.

There are several problems associated with recrystallization:

— In softeners that use a float to control the amount of water for brine composition, the recrystallized KCl will occupy space normally occupied by the brine. The lower volume will decrease the amount of KCl used in regeneration, which will reduce the expected capacity of the softener.
— In all softeners, if the recrystallization is left unchecked, KCl deposits can plug the brine tank valve, causing the softener to malfunction.

There is one known solution for the problem of recrystallization: the temperature of the brine tank must be kept constant. Built-in heaters that maintain constant temperature could be installed by equipment manufacturers to serve this function. In most cases, however, the recrystallization problem can be managed, or eliminated by the following practices:

— Placing the brine tank on a four or six inch wooden platform.
— Placing a piece of 2-inch styrofoam under the brine tank.
— Insulating the brine tank to a height just above the brine level.
— Removing the grid plate from the brine tank.
— When the regenerant has been depleted, using a broom handle to break up any recrystallization that has occurred on the bottom of the tank and then refilling the brine tank to approximately half of its capacity with fresh KCl regenerant.

Research has also been done to explore the possibility of adding reagents to prevent or slow down the process of recrystallization. The reagents examined included a crystal growth modifier, potassium acetate, potassium pyrophosphate, potassium phosphate monobasic, and a metal retardant. None of these reagents affected the bridging (recrystallization) problem.

Research may one day find a way to eliminate the recrystallization problem with KCl completely, but for now the precautions listed above are the best solution. Most customers have found that the above precautions are adequate to manage or solve the recrystallization problem with KCl in their softeners. Some people have found no problems at all.

FREQUENTLY ASKED QUESTIONS

To Convert from NaCl to KCl Regenerant, Do I Need to Make any Changes to the Softener Equipment or Settings?

No changes to the softener equipment are required for the conversion to KCl regenerant. The softener settings need only be adjusted if it is a high-efficiency unit; in such a case the regenerant level is typically set at 4 pounds per cubic foot of resin, and when converting to KCl the setting should be increased to 4.5 pounds per cubic foot. At higher regenerant dosages, there is no need to change settings when converting to KCl. The reasons why KCl provides softening equivalent to NaCl are discussed in detail in Chapter 8.

Does My Tank Need to be Empty to Switch to KCl Regenerant?

No, there is no incompatibility between KCl and NaCl regenerants. Potassium chloride can be simply poured into an existing softener that already contains NaCl.

What Can be Done to Prevent Bridging of KCl Regenerant?

Like NaCl, the use of KCl regenerant can lead to bridging. However, the source of the problem (and the symptoms) with KCl are different than with NaCl. Bridging can occur when the KCl regenerant tank is exposed to temperature fluctuations, resulting in a layer of recrystallized KCl on the bottom of the tank. Several practical steps can be taken to reduce or eliminate this recrystallization, as discussed in detail in the preceding section.

Can KCl-Softened Water be Used for Watering House Plants?

Yes, water softened with KCl can be used for house plants provided one follows the normal good practice of allowing excess water to drain from the bottom of the pot. Water softened with KCl can be a good source of potassium, which is an essential plant nutrient.

Can KCl-Softened Water be Used in My Fish Tank?

No, the use of any softened water (regenerated with either NaCl or KCl) is not recommended for use in fish tanks.

Can KCl-Softened Water be Given to Pets?

Yes, potassium-softened water is safe for pets. The softened water will contain a small, but significant, amount of potassium, which is an important nutrient for both humans and pets.

Can KCl-Softened Water be Used for Watering My Lawn and Garden?

Yes, KCl-softened water is safe to use in residential watering applications. In fact, potassium is a primary plant nutrient, so plants will acquire some nutrient values from the softened water. It is a good idea, however, to alternate between watering with softened and hard water, so that the plants receive a good balance of potassium, calcium, and magnesium.

Can KCl Regenerant Waste be Used for Watering My Lawn and Garden?

The regenerant waste from a KCl softener is a good source of potassium. However, the regenerant brine is far too concentrated for residential watering. Waste regenerant can be used in appropriate agricultural settings, where the waste brine can be diluted, and there is adequate land available. This point is discussed in more detail in Chapter 9.

How Efficient is KCl at Removing Iron and Manganese?

Home softeners are often able to remove low levels of iron and manganese. Tests have shown that the performance of KCl is comparable to that of NaCl with respect to removal of these elements. This point is discussed in more detail in Chapter 11, in the section "Iron and Manganese Removal."

Will KCl Soften My Water as Effectively as NaCl?

Yes, conversion from NaCl to KCl will not have any impact on hardness reduction. Problems with scale formation, soap curd, and dry itchy skin will be eliminated by softening with either regenerant.

Are There any Medical Issues Associated with Conversion to KCl Regenerant?

Conversion to KCl is normally a healthy choice for most people, since it results in greater potassium and lower sodium intake in the drinking water. Potassium is an essential nutrient, while most people's daily sodium intake is too high. However, a small but significant number of people are at risk of potassium overload, as discussed in Chapter 10. If people are concerned about their potassium intake, we recommend that they consult a medical professional.

REFERENCES

Clescerl LS, Greenberg AE, Eaton AD, editors. *Standard Methods for the Examination of Water and Wastewater*. American Public Health Association; 20th edition; January 1999.

McGowan W. *Water Processing for Home, Farm and Business*. Cole L, technical editor. Naperville, IL: Water Quality Association; 1988.

Weast RC, editor. *CRC Handbook of Chemistry and Physics*. CRC Press; 1977.

CHAPTER 8

COMPARISON OF KCI AND NaCl AS REGENERANT

Sodium chloride is very well established as the standard regenerant in water softening. Therefore, if potassium chloride is to be accepted as a good alternative to sodium chloride, it is necessary to provide a comparison between the performance of these two regenerants. In the past several years, numerous tests have been performed to compare the performance of KCl and NaCl as regenerants in water softening. The results of these tests are presented in this chapter. Before the comparisons, however, we need to define some terms to quantify the performance of water softeners.

DEFINITION OF TERMS

The water treatment industry has defined many terms so that a fair quantitative comparison can be made between different softeners, or between softeners supplied by different manufacturers. In the present text we are most interested in the comparison in performance between KCl and NaCl as regenerant. We therefore restrict our focus to those terms necessary to compare KCl and NaCl.

Water Softening with Potassium Chloride: Process, Health, and Environmental Benefits,
by William Wist, Jay H. Lehr, and Rod McEachern
Copyright © 2009 by John Wiley & Sons, Inc.

Grain

The grain is a unit of weight (technically speaking, mass) equal to 0.0648 grams. In the British system, one grain is equal to 1/7000 of a pound (0.000143 pounds).

Grains per Gallon

Water hardness is often reported in North America in terms of grains per US Gallon (gpg). Most commonly, the reported hardness will be in terms of gpg expressed in terms of calcium carbonate equivalent. One grain per gallon hardness is equivalent to 17.1 mg/L (ppm).

Soft Water

Soft water is defined as water (naturally occurring) that contains less than 1.0 grains per gallon of total hardness, expressed in terms of the calcium carbonate equivalent. Hardness of 1.0 grains is equivalent to 17.1 mg/L (ppm) calcium carbonate equivalent.

Softened Water

Softened water is water, initially hard, that has been treated by ion exchange or other process so that the total hardness has been reduced to 1.0 grains per gallon or less.

Regeneration

Regeneration is the process of flushing a solution (typically an NaCl or KCl brine) through an ion exchange resin so that the contaminant ions (Ca^{2+}, Mg^{2+}, and others) are removed and replaced with the desired ions (Na^+ or K^+). Regeneration restores the capacity of the resin and makes it ready for reuse.

Regeneration Cycle

The regeneration cycle is the sequence of steps (rinse, backflush, brine draw, etc., as described in detail in Chapter 6) in which a solution is used to regenerate a resin and make it ready for reuse.

Regenerant Level

The regenerant level is the quantity of regenerant (typically NaCl or KCl) used in the regeneration cycle of a water softener. The regenerant level is typically expressed in terms of pounds of regenerant per cubic foot of ion exchange resin. Alternately, for a given ion exchange system, the regenerant level may be expressed simply in terms of pounds of regenerant. The regenerant level is also known as the salt dosage.

Rated Service Flow

The rated service flow is the range (minimum and maximum) flows at which a water softener can be operated while continuously producing soft water. The rated service flow is defined by the manufacturer of the softening equipment.

Rated Softener Capacity

The rated softener capacity is one of the key parameters describing the effectiveness of a water softening system. There are several different ways in which manufacturers can express the rated water softener capacity of their products; all such methods are a quantitative measure of the amount of water that can be softened by the unit:

— Capacity can be expressed in terms of the expected number of days the unit will be in service between regeneration cycles,
— It can be expressed in terms of the expected gallons of water treated between regeneration cycles, or
— It can be expressed in terms of the total grains hardness removed between regeneration cycles.

Regardless of the choice of method to describe rated softener capacity, the reported capacity will be a function of the regenerant level used, so this must be specified as part of the statement of capacity. For example, if a water softener is treating water and the following performance is observed:

— The influent hardness is 32.4 gpg
— 525 gallons is softened before the effluent hardness rises to 1.0 gpg
— The regenerant level is 4 pounds of sodium chloride

then we would state that the rated softener capacity for this system is:

$$525 \text{ gallons} \times 32.4 \text{ gpg} = 17{,}010 \text{ grains}$$

at a regenerant level of 4 pounds of sodium chloride.

THEORETICAL CAPACITIES OF KCl AND NaCl REGENERANTS

We know from Chapter 4 that two ions of sodium (Na^+) or potassium (K^+) can displace one ion of calcium (Ca^{2+}) or magnesium (Mg^{2+}) from an ion exchange resin, and thereby regenerate the resin for service. The molecular weights of KCl and NaCl are, however, different (Chapter 5), so one would expect that the capacity for regeneration would not be the same for these two substances on a per pound basis. We can calculate the theoretical capacity for a pound of NaCl as follows.

$$1 \text{ pound NaCl} \times \frac{453.59 \text{ grams}}{\text{pound}} = 453.9 \text{ grams}$$

We know from Chapter 5 that 1 mole of sodium chloride has a mass of 58.443 grams. Therefore, we can calculate the number of moles of sodium chloride in 1 pound:

$$\frac{453.9 \text{ grams}}{58.443 \dfrac{\text{grams}}{\text{mole}}} = 7.7665 \text{ moles}$$

We know that during regeneration the reaction that occurs to displace calcium ions from the ion exchange resin will be (Chapter 3):

$$2R\text{--}SO_3^-Na^+ + Ca^{2+} \rightleftharpoons (R\text{--}SO_3^-)_2 \, Ca^{2+} + 2Na^+$$

with the reaction proceeding to the left during regeneration. Therefore, 2 moles of sodium ions will be required to displace each mole of calcium ions. Therefore, 7.7665 moles of sodium chloride will displace $7.7665 \div 2 = 3.8833$ moles of calcium ions. We know that the molecular weight of calcium carbonate is 100.088 grams/mole (Table 5-2), so we can express 3.8833 moles of calcium on a calcium carbonate equivalence basis:

$$3.8833 \text{ moles} \times 100.088 \frac{\text{grams CaCO}_3}{\text{mole}} = 388.67 \text{ grams CaCO}_3$$

Finally, we know that one grain is equal to 0.064799 grams (Table 3-2), so we can convert the calcium displaced from the resin to grains, expressed as $CaCO_3$:

$$388.67 \text{ grams CaCO}_3 \times \frac{1 \text{ grain}}{0.064799 \text{ grams}} = 5998 \text{ grains}$$

The above analysis indicates that in theory (that is, if 100% efficiency is achieved) 1 pound (453.9 grams) of sodium chloride would be able to displace 5998 grains of calcium, expressed as calcium carbonate. We would therefore state that the theoretical regeneration capacity of sodium chloride is 5998 grains (as $CaCO_3$) per pound.

The theoretical regeneration capacity for potassium chloride can be calculated similarly. The mass of 1 pound (453.9 grams) of potassium chloride will be:

$$\frac{453.9 \text{ grams}}{74.551 \dfrac{\text{grams}}{\text{mole}}} = 6.0885 \text{ moles}$$

The reaction in which potassium displaces calcium from the ion exchange resin is similar to that for sodium:

$$2R\text{–}SO_3^-K^+ + Ca^{2+} \rightleftharpoons (R\text{–}SO_3^-)_2\, Ca^{2+} + 2K^+$$

so 6.0885 moles of potassium will displace $6.0885 \div 2 = 3.0442$ moles of calcium. Conversion to grams of calcium carbonate is then:

$$3.0442 \text{ moles} \times 100.088 \frac{\text{grams CaCO}_3}{\text{mole}} = 304.69 \text{ grams CaCO}_3$$

and finally, conversion to grains of $CaCO_3$ gives:

$$304.69 \text{ grams CaCO}_3 \times \frac{1 \text{ grain}}{0.064799 \text{ grams}} = 4702 \text{ grains}$$

The above analysis shows that on a per-pound basis, the theoretical regeneration capacity for sodium chloride is significantly higher than that of potassium chloride (5998 grains vs. 4702 grains, expressed on a $CaCO_3$ basis). However, as we shall see later in this chapter, other factors come into play that make the performance of KCl comparable to that of NaCl.

CALCULATION OF REGENERATION EFFICIENCY

There are various methods of calculating regeneration efficiency. One of the most important is the ion exchange resin capacity divided by the pounds of regenerant used; this efficiency calculation is also known as the *salt efficiency*. For example, if a softener has a capacity of 15,000 grains as $CaCO_3$ when regenerated with 4.00 pounds of sodium chloride, then we can calculate the efficiency:

$$\text{Salt efficiency} = \frac{15,000 \text{ grains}}{4.00 \text{ pounds}} = 3750 \text{ grains/pound}$$

If the softener described above was regenerated with 8.00 pounds of sodium chloride rather than 4.00 pounds then one would find that the capacity would increase, but not in proportion to the amount of salt used. For example, regeneration of the above ion exchange resin with 8.00 pounds of sodium chloride would perhaps increase the capacity to 20,000 grains of hardness. With 8.00 pounds of regenerant, the efficiency would then be:

$$\text{Salt efficiency} = \frac{20,000 \text{ grains}}{8.00 \text{ pounds}} = 2500 \text{ grains/pound}$$

We see that increasing the regenerant dosage increases the softener capacity, but at the cost of decreased salt efficiency. There is thus a trade-off between softener capacity (with its associated cost) and efficiency (with the associated costs of salt usage).

Rather than reporting the efficiency of a regenerant in terms of the salt efficiency, we can calculate the regenerant efficiency in relation to the theoretical capacity. We define this term as the *specific efficiency*:

$$\text{Specific efficiency} = \frac{\text{Actual capacity as } CaCO_3}{\text{Theoretical capacity}} \times 100\%$$

For example, we saw in the preceding section that 1 pound of NaCl has the theoretical ability to displace 5998 grains of hardness (as $CaCO_3$) from an ion exchange resin. Therefore, if the above ion exchange system with 15,000 grains capacity was regenerated with 4.00 pounds of NaCl regenerant, then the specific efficiency could be calculated:

$$\text{Specific efficiency} = \frac{15{,}000 \text{ grains}}{4 \text{ pounds} \times 5998 \dfrac{\text{grains}}{\text{pound}}} \times 100\% = 62.5\%$$

Similarly, we know that the theoretical capacity of potassium chloride is 4702 grains hardness per pound. Therefore if the softening system achieved a capacity of 15,000 grains when regenerated with potassium chloride, then the specific efficiency would be:

$$\text{Specific efficiency} = \frac{15{,}000 \text{ grains}}{4 \text{ pounds} \times 4702 \dfrac{\text{grains}}{\text{pound}}} \times 100\% = 79.8\%$$

Similarly, in the case where 8.00 pounds of regenerant achieved a capacity of 20,000 grains, we would have a specific efficiency for sodium chloride:

$$\text{Specific efficiency} = \frac{20{,}000 \text{ grains}}{8 \text{ pounds} \times 5998 \dfrac{\text{grains}}{\text{pound}}} \times 100\% = 41.7\%$$

or for potassium chloride:

$$\text{Specific efficiency} = \frac{20{,}000 \text{ grains}}{8 \text{ pounds} \times 4702 \dfrac{\text{grains}}{\text{pound}}} \times 100\% = 53.2\%$$

Both the salt efficiency and the specific efficiency can be useful ways to describe the performance of a regenerant, and we shall use both in the discussion that follows later in this chapter. We emphasize that the performance numbers given above are examples, for illustration only, and do not provide an actual comparison of the performance of KCl relative to NaCl.

SIZING A SOFTENER FOR SALT EFFICIENCY

In Chapter 6 we illustrated how the capacity requirements of a water softener can be calculated. In the example from that chapter, we considered a typical household installation, with the following characteristics:

— Four occupants with a daily consumption of 300 gallons
— Influent hardness of 34 grains per gallon
— Daily demand of 10,200 grains
— 4 days between regeneration cycles
— Softener capacity of 40,800 grains

As mentioned in Chapter 6, proper sizing requires consideration of more than just ion exchange capacity. In practice, the flow rate of water through the softener is an important factor that must be considered. The flow rate that the softener experiences will be a function of the number of fixtures in the house, as well as the details of the household plumbing. Intuitively, we can understand why the flow rate will impact the required volume of ion exchange resin, because a very large flow through a very small resin bed will not allow enough time for the softening reaction to occur. Suppliers provide water softeners with a maximum rated service flow, which ensures that there is an adequate bed volume to properly soften influent water at the maximum flow. The relationship between rated service flow and the required volume of ion exchange resin is illustrated in Figure 8-1 for a typical softener.

From Figure 8-1 we can see that if our typical household had an expected service flow of 15 gallons per minute, then an ion exchange bed of 2.2 ft³ would be recommended. To ensure adequate softening at times of high demand, it would be prudent to round up, and install a system with 2.5 ft³ of ion exchange resin bed.

For any given size of resin bed, the capacity will vary with the amount of salt regenerant used, since larger amounts of salt will push

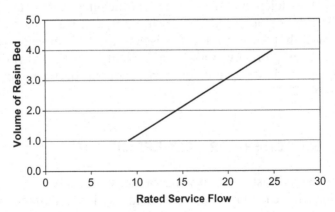

FIGURE 8-1 Required volume of resin bed (in ft³) as a function of the rated service flow (gpm) for a typical water softening application.

TABLE 8-1: Water softener capacity (grains) as a function of the volume of resin (ft^3) and the salt dosage (pounds per cubic foot of resin)

Volume of Ion Exchange Resin (ft^3)	Salt Dosage (pounds per cubic foot of resin)			
	6	8	10	15
1.0	20,000	24,000	27,000	30,000
1.5	30,000	36,000	40,500	45,000
2.0	40,000	48,000	54,000	60,000
2.5	50,000	60,000	67,500	75,000
3.0	60,000	72,000	81,000	90,000
4.0	80,000	96,000	108,000	120,000
5.0	100,000	120,000	135,000	150,000

the regeneration reaction further and thereby utilize a greater amount of the resin's inherent capacity. However, the use of larger amounts of salt comes at a cost—higher salt usage results in lower salt efficiency and therefore higher overall salt costs. Data for typical ion exchange resin capacity as a function of resin volume and salt dosage are given in Table 8-1.

Table 8-1 can be used to complete the sizing for a water softener for our example household. We saw that 2.5 cubic feet of resin bed would be required for the expected service flow. We also know that in this application, 40,800 grains of capacity will be required. Therefore, in Table 8-1 we look across the row for 2.5 ft^3 bed volume to determine the salt dosage required to achieve 40,800 grains of capacity. We see from Table 8-1 that 6 pounds of salt per cubic foot of resin will provide more than adequate capacity.

The final specifications for our example household will therefore be:

— 40,800 grains capacity
— 2.5 cubic feet bed volume
— 6 pounds salt per cubic foot of resin, per regeneration cycle.

IMPLICATIONS FOR SALT CONSUMPTION

Careful examination of Table 8-1 provides insight into salt efficiencies in water softeners. For any given bed volume, as we proceed right from 6 lbs per ft^3 to 15 lbs per ft^3 we see that the capacity of the softener rises. However, the capacity does not rise in proportion to the salt dosage, which is an indication of less efficient use of salt at higher dosages.

When a softener is regenerated, the softening reaction is shifted to the left, returning the resin to the $R-SO_3^-$ Na^+ form:

$$2R-SO_3^-Na^+ + Ca^{2+} \rightleftharpoons (R-SO_3^-)_2 Ca^{2+} + 2Na^+$$

In practice, no one regenerates their household softener 100% into the $R-SO_3^-$ Na^+ form. The regeneration reaction is controlled by chemical equilibrium; as a result the first salt to contact the resin during regeneration will be very effective at removing adsorbed hardness. However, as regeneration proceeds further the sodium ions in the brine will be less and less effective, that is, the regeneration process has diminishing returns for addition of extra sodium.

Regeneration with progressively larger salt dosages will indeed increase the capacity of the resin because a larger fraction of the ions on the resin will be sodium at the start of the service cycle. However, salt efficiency will decrease as the salt dosage increases, because of the diminishing returns for the extra sodium. Consider, for example, the data for a bed volume of $2.5\,ft^3$ in Table 8-1. We see that the resin capacity increases from 50,000 to 75,000 grains on going from 6 to 15 pounds salt per ft^3. From the section "Calculation of Regeneration Efficiency" we know that the salt efficiency can be calculated:

$$\text{Salt efficiency} = \frac{\text{Resin capacity in grains}}{\text{pounds of regenerant salt}}$$

In the case of regeneration with 6 lbs of salt per ft^3, we achieve a capacity of 50,000 grains. Given that the bed volume is $2.5\,ft^3$ the total dosage of salt will be:

$$\text{Salt Dosage} = 6\,\text{pounds}/ft^3 \times 2.5\,ft^3 = 15.0\,\text{pounds}$$

We can therefore calculate the salt efficiency:

$$\text{Salt efficiency} = \frac{50,000\,\text{grains}}{15.0\,\text{pounds}} = 3333\,\text{grains/pound}$$

In contrast, regeneration with 15 pounds of salt per ft^3 will require a total salt dosage of:

$$\text{Salt Dosage} = 15\,\text{pounds}/ft^3 \times 2.5\,ft^3 = 37.5\,\text{pounds}$$

and the reported capacity of 75,000 grains gives a salt efficiency of:

$$\text{Salt efficiency} = \frac{75{,}000 \text{ grains}}{37.5 \text{ pounds}} = 2000 \text{ grains/pound}$$

In a similar manner, the salt efficiencies were calculated for all bed volumes and salt dosages shown in Table 8-1. Note that the salt dosage is scaled up in proportion to the bed volume, so the salt efficiency does not depend on the bed volume. For all bed volumes, the salt efficiency varies with salt dosage as shown in Table 8-2.

The impact of increasing salt dosage on both capacity and salt efficiency is illustrated in Figure 8-2.

Figure 8-2 illustrates the trade-off between softener capacity and salt efficiency; increasing the salt dosage will increase the capacity of a given softener, but the decrease in efficiency will result in higher salt usages, which comes at a cost (to both your pocket book and the environment).

TABLE 8-2: Calculated salt efficiencies (grains/pound) as a function of the salt dosage (pounds per cubic foot of resin) for the data shown in Table 8-1

	Salt Dosage (pounds per cubic foot of resin)			
	6	8	10	15
Salt Efficiency	3333	3000	2700	2000

Note that the salt dosage was scaled up in proportion to the bed volume so the salt efficiency is independent of bed volume.

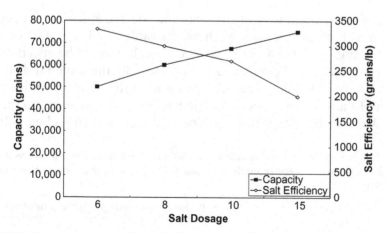

FIGURE 8-2   Impact of salt dosage (in pounds per cubic foot of bed volume) on the softener's capacity (grains) and salt efficiency (grains per pound of salt). Data in this figure have been calculated for a bed volume of $2.5\,\text{ft}^3$, which impacts calculated capacity but not salt efficiency (see text).

We saw in the section "Theoretical Capacities of KCl and NaCl Regenerants" that the efficiency of a softening system can also be reported in terms of the theoretical capacity for regeneration of each pound of salt. One pound of NaCl has the theoretical capacity to remove 5998 grains of hardness from an ion exchange resin. Thus when 6 pounds per ft^3 of NaCl regenerant is added to an ion exchange system with a bed volume of $2.5\,ft^3$ it has the capacity to remove hardness:

$$\text{Capacity} = 2.5\ ft^3 \times 6\ \text{pounds}/ft^3 \times 5998\ \text{grains}/\text{pound}$$

$$\text{Capacity} = 89,820\ \text{grains}$$

From Table 8-1 we saw that regeneration of a softener with a bed volume of $2.5\,ft^3$ with 6 pounds per ft^3 of NaCl yielded a capacity of 50,000 grains. Therefore, we can calculate the specific efficiency (see the section "Calculation of Regenerant Efficiency") for this system:

$$\text{Specific efficiency} = \frac{\text{Actual capacity as } CaCO_3}{\text{Theoretical capacity}} \times 100\%$$

in this case:

$$\text{Specific efficiency} = \frac{50,000\ \text{grains}}{6\ \text{pounds}/ft^3 \times 2.5\ ft^3 \times 5998\ \dfrac{\text{grains}}{\text{pound}}} \times 100\% = 55.6\%$$

In a similar manner, we can calculate the specific efficiency for regeneration with other salt dosages shown in Table 8-1. The results of such a calculation are shown in Table 8-3. Calculation of the specific efficiency may seem to be redundant, as essentially the same information is provided as by doing a calculation on the salt efficiency (Table 8-2). The usefulness of the specific efficiency lies in the ease of interpretation. We saw above that regeneration of a softener with a bed volume

TABLE 8-3: Calculated specific efficiency (%) and percentage unused salt, as a function of the salt dosage (pounds per cubic foot of resin) for the data shown in Table 8-1

	Salt Dosage (pounds per cubic foot of resin)			
	6	8	10	15
Specific efficiency (%)	55.6	50.0	45.0	33.3
Percent unused salt (%)	44.4	50.0	55.0	66.7

Note that the salt dosage was scaled up in proportion to the bed volume so the salt efficiency is independent of bed volume.

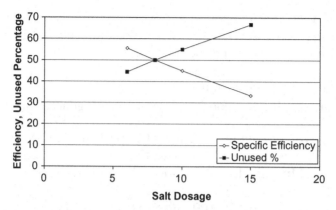

FIGURE 8-3   Impact of salt dosage (in pounds per cubic foot of bed volume) on the specific efficiency and the unused percentage for the salt regenerant. Data in this figure have been calculated for a bed volume of 2.5 ft³, which impacts calculated capacity but not salt efficiency (see text).

of 2.5 ft³ at a salt dosage of 6 pounds per ft³ gives a specific efficiency of 55.6%. What does this number mean? The specific efficiency tells us the fraction of the salt that actually is used to help regenerate the resin; in the present example, 55.6% of the sodium ions in the regenerant solution displace hardness ions from the resin and become adsorbed themselves. The remainder (44.4%) of the sodium ions are unused; they simply pass through the ion exchange bed and report to waste. The specific efficiency and the percentage unused are therefore obvious indications of how much value is being achieved for each dollar spent on salt. In addition, the environmental costs of low efficiency are significant since the unused salt has an environmental cost through increased sodium and chloride loading to the environment, with no benefits achieved.

The specific efficiency and the percentage unused salt are illustrated in Figure 8-3 as a function of salt dosage.

TOTAL SALT RELEASED TO THE ENVIRONMENT

In Chapter 6 we examined a "typical" household with the following water requirements:

— Four occupants with a daily consumption of 300 gallons
— Influent hardness of 34 grains per gallon
— Daily demand of 10,200 grains
— 4 days between regeneration cycles

and determined that the softener would require a capacity of 40,800 grains. In the above sections we then calculated the bed volume and salt requirements and found them to be:

— 2.5 cubic feet bed volume
— 6 pounds salt per cubic foot of resin, per regeneration cycle

Since we have completely described the performance of this typical household softener, we can calculate the amount of sodium chloride used per year. In a single regeneration, the unit will use:

$$2.5 \, \text{ft}^3 \text{ resin} \times 6 \, \frac{\text{pounds salt}}{\text{ft}^3 \text{ resin}} = 15 \text{ pounds salt}$$

The softener regenerates once every 4 days; therefore in a year the total amount of regenerant used will be:

$$15 \text{ pounds salt/cycle} \times \frac{365 \text{ days/year}}{4 \text{ days/cycle}} = 1369 \text{ pounds/year}$$

We see that the total salt consumption per year is therefore 1369 pounds. The amount of salt used in a typical household is substantial, both in terms of cost for the purchase of the salt, as well as the hassle in keeping the softener filled with salt. Moreover, all of the salt added to the softener will eventually make its way into the environment which (if sodium chloride is used as regenerant) increases the loading of chloride and sodium, both of which can be harmful to the environment.

We saw in Table 8-3 that 44.4% of the salt used in the regeneration cycle was unused; that is, this portion of the salt was not adsorbed on the resin during regeneration. Instead, this 44.4% of the regenerant salt simply passes through the resin bed and moves directly into the environment. Over the course of a year, the total amount of unused salt contributed to the environment will be:

$$1369 \text{ pounds} \times \frac{44.4\%}{100} = 608 \text{ pounds}$$

This salt increases the loading of sodium and chloride to the environment without providing any real benefits to the softening process. The specific efficiency of a softening system is therefore an important parameter when considering the environmental impact of softening.

The key concepts were introduced, and key terms defined in the preceding sections. We are therefore now in a position to provide a detailed comparison between the performance of KCl and NaCl regenerants. Such comparisons are provided in the remainder of this chapter.

COMPARISON OF KCl AND NaCl: SOLUBILITY

Regeneration of an ion exchange resin occurs when the softening reaction:

$$2R\text{–}SO_3^-M^+ + Ca^{2+} \rightleftharpoons (R\text{–}SO_3^-)_2\, Ca^{2+} + 2M^+ \qquad (M = Na \text{ or } K)$$

is reversed, that is, shifted to the left. The driving force for this reaction is the large excess of Na^+ or K^+ ions in the regenerant solution. It makes sense, then, that a regenerant solution will be more effective if the regenerant salt has a high solubility in water.

Both NaCl and KCl are effective regenerants for softening, in part because they have very similar solubility behavior at room temperature, as shown in Table 8-4 and Figures 8-4 and 8-5. As the temperature is raised, the solubility of KCl increases substantially while NaCl

TABLE 8-4: Solubility of KCl and NaCl as a function of temperature, expressed in various units

	Temperature (°C)										
	0	10	20	30	40	50	60	70	80	90	100
Grams in 100 grams of water											
NaCl	35.6	35.7	35.8	36.0	36.3	36.7	37.1	37.5	38.0	38.5	39.1
KCl	27.6	31.0	34.0	37.0	40.0	42.6	45.5	48.3	51.1	54.0	56.7
Grams in 100 grams of a saturated solution											
NaCl	26.3	26.3	26.4	26.5	26.6	26.8	27.1	27.3	27.5	27.8	28.1
KCl	21.6	23.7	25.4	27.0	28.6	29.9	31.3	32.6	33.8	35.1	36.2
Pounds in 1 US gallon of water											
NaCl	3.0	3.0	3.0	3.0	3.0	3.1	3.1	3.1	3.2	3.2	3.3
KCl	2.3	2.6	2.8	3.1	3.3	3.5	3.8	4.0	4.3	4.5	4.7
Pounds in 1 US gallon of saturated brine (SG = 1.20)											
NaCl	2.6	2.6	2.6	2.7	2.7	2.7	2.7	2.7	2.8	2.8	2.8
KCl	2.2	2.4	2.5	2.7	2.9	3.0	3.1	3.3	3.4	3.5	3.6

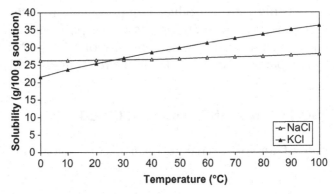

FIGURE 8-4   Solubility of KCl and NaCl (in grams of salt per 100 grams of a saturated solution) as a function of temperature in degrees Celsius.

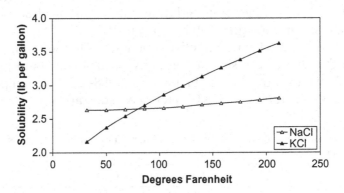

FIGURE 8-5   Solubility of KCl and NaCl (in pounds of salt per gallon of saturated solution) as a function of temperature in degrees Fahrenheit.

solubility increases only modestly. However, the vast majority of softening applications are near room temperature, and under these conditions the difference in solubility between KCl and NaCl is very slight, thus the impact of solubility will favor neither KCl nor NaCl.

COMPARISON OF KCI AND NaCI: SPEED OF DISSOLUTION

Water softeners that regenerate on a fixed time and fixed salt level add water to the salt brine tank during the regeneration cycle. In this case, the water typically has several days to reach saturation, so the speed of dissolution is not a concern.

TABLE 8-5: Extent of saturation (in weight percent) of NaCl and KCl as a function of time and percentage of saturation for the resulting brine

Time (min.)	NaCl		KCl	
	Wt. % in Sol'n	% Saturated	Wt. % in Sol'n	% Saturated
5	23.3	88.6	20.3	89.0
10	25.1	95.4	20.9	91.7
15	26.0	98.9	22.1	96.9
30	26.3	100	22.7	99.6

The new computerized water softeners now use a different system of brine make-up. Regeneration only occurs on demand. The computer uses information such as hardness and the volume of water treated to determine when regeneration is required. A calculation on the amount of salt required is done just before regeneration. The water for brine make-up is then added and regeneration begins almost immediately. This means that in a matter of minutes the brine is being eluted. In such a case, there is a potential problem because the salt requires a certain amount of time for dissolution, to form the regenerant solution.

A series of tests were therefore performed to determine whether the rate of dissolution of KCl was comparable to that of NaCl under water softener conditions. A 1000-ml graduated cylinder was filled with KCl, and a second cylinder was filled with NaCl. Then, using a graduated cylinder for measurement, 250 ml of tap water at 6 °C was poured into each salt cylinder. After a specified time, the resulting brine was poured out of the salt cylinder and analyzed. Typical results from this series of tests are shown in Table 8-5 and illustrated in Figure 8-6. In addition to the solubility as a function of time, Table 8-5 provides the degree of saturation for each solution, that is, the salt concentration (in weight percent) divided by the salt concentration of a saturated solution (also in weight percent) and expressed as a percentage. For reference, a saturated solution of KCl is 22.8% at 6 °C, while a saturated solution of NaCl is 26.3% at the same temperature.

In the above set of tests, 6 °C was chosen as the temperature for the tap water, so that the test represented a worst-case scenario. That is, it is doubtful that the water used for making the regenerant solution would be any colder than this in a household water softening system.

The testing showed that the rate of dissolution is very rapid for both KCl and NaCl provided that the make-up water is totally in contact

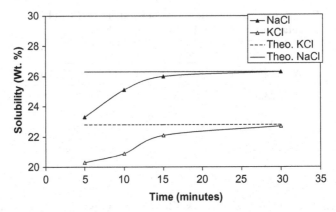

FIGURE 8-6   Extent of saturation of NaCl and KCl as a function of time. Also included is a line indicating the concentration of saturated solutions.

with the salt. A near-saturated solution is formed within 12–15 minutes, and dissolution is essentially complete within 30 minutes. As a result, we conclude that the performance of KCl and NaCl are comparable. There would be no performance differences when switching from NaCl to KCl for most softening systems. Regeneration on demand systems may require a minor change to the computer data when switching from one salt to another.

COMPARISON OF KCl AND NaCl: CAPACITY

The capacity of KCl, relative to NaCl, for water softening is one of the key issues that must be addressed to show that KCl is a viable alternative to NaCl. We saw in Chapter 7 that potassium chloride (KCl) has a molecular weight of 74.55 g/mole, in contrast to the value of 58.43 g/mole for sodium chloride (NaCl). The molecular weight of KCl is therefore 27.6% higher than that of NaCl. On the basis of simple chemical principles we would then expect that pound for pound, 27.6% more KCl would be required than NaCl. This simple picture, however, is not observed in practice. Experimental results described in this chapter (and shown in detail in Appendixes 1 and 2) show that the capacity of KCl for softening is comparable to that of NaCl, except at very low salt dosages.

A detailed and standardized test protocol was developed (Appendix 1) to compare KCl and NaCl as regenerants. The test protocol

TABLE 8-6: Average test results for NaCl regenerant by the standardized test protocol

Regeneration Pounds	U.S. Gallons per Cycle	Influent Hardness as $CaCO_3$	Grains Exchanged per ft³, as $CaCO_3$	System Efficiency grains/ft³/pound
4	542	30.4	16,408	4102
8	850	30.4	25,585	3189
15	1,098	30.4	33,278	2219

Detailed experimental results for each of the four laboratories are provided in Appendix 2.

TABLE 8-7: Average test results for KCl regenerant by the standardized test protocol

Regeneration Pounds	U.S. Gallons per Cycle	Influent Hardness as $CaCO_3$	Grains Exchanged per ft³, as $CaCO_3$	System Efficiency grains/ft³/pound
4	482	30.3	14,505	3626
8	833	30.4	25,255	3157
15	1,067	30.6	32,552	2170

Detailed experimental results for each of the four laboratories are provided in Appendix 2.

was followed by four laboratories independently. The detailed experimental results from each of the four laboratories are given in Appendix 2. In the current section, we will discuss and analyze the average of the test results; the average data are presented in Tables 8-6 and 8-7.

Comparison of Tables 8-6 and 8-7 shows that the capacity for softening (grains hardness exchanged per cubic foot of resin) is very similar between KCl and NaCl. The comparison in capacity (grains exchanged per cubic foot, as $CaCO_3$) between KCl and NaCl is illustrated in Figure 8-7.

Figure 8-7 shows that the capacity of NaCl as a regenerant is slightly higher than that of KCl only at the lowest regenerant dosage (4 pounds). At the higher regenerant dosages (8 and 15 pounds) the differences are very slight, and not significant. We can quantify these statements by calculating the ratio of the capacity of NaCl to that of KCl. With four pounds of regenerant we have the capacities:

$$NaCl \text{ capacity} = 16,408 \text{ grains (as } CaCO_3) \text{ per ft}^3 \text{ of resin}$$

$$KCl \text{ capacity} = 14,505 \text{ grains (as } CaCO_3) \text{ per ft}^3 \text{ of resin}$$

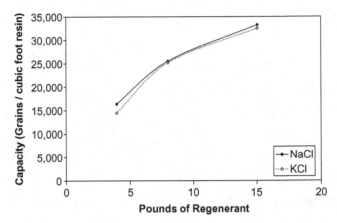

FIGURE 8-7 Capacity for softening (expressed as grains of hardness as $CaCO_3$ per cubic foot of ion exchange resin) for KCl and NaCl at various regenerant dosages. The data shown in this figure are from Tables 8-6 and 8-7 and were obtained with the standardized test protocol.

TABLE 8-8: NaCl:KCl capacity ratio (grains of hardness removed as $CaCO_3$ per cubic foot of ion exchange resin) for various regenerant dosages

Pounds of Regenerant	Capacity Ratio
4	1.131
8	1.013
15	1.022

Results are calculated with the data shown in Tables 8-6 and 8-7.

The ratio of capacities is then:

$$\text{Capacity ratio} = \frac{\text{NaCl capacity}}{\text{KCl capacity}} = \frac{16,408 \text{ grains}}{14,505 \text{ grains}} = 1.131$$

Similarly, we can calculate the capacity ratio for other regenerant dosages, using the data from Tables 8-6 and 8-7. The results are shown in Table 8-8 and illustrated in Figure 8-8.

The data provided in Table 8-8 and illustrated in Figure 8-8 show that NaCl provides a modest (13%) improvement in capacity relative to KCl, only at the lowest regenerant dosage. Above 4 pounds of regenerant there are no significant differences between the capacity of NaCl and that of KCl. Moreover, if one were to convert from NaCl to KCl

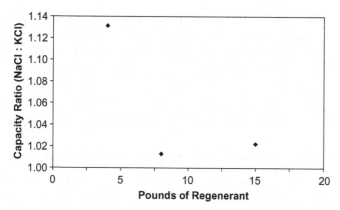

FIGURE 8-8 NaCl:KCl capacity ratio (grains of hardness removed as $CaCO_3$ per cubic foot of ion exchange resin) as a function of regenerant dosage.

at the 4 pounds of regenerant level, the increased amount of regenerant required would be modest—4 lbs. × 1.131 = 4.524 lbs.

The data shown in Table 8-8 are somewhat surprising. Examination of the molecular weights, at the start of this section, showed that in theory 27.6% more regenerant would be required when converting from NaCl to KCl. Otherwise, the capacity of the softening system would suffer. The results shown in Table 8-8 indicate that the theoretical results do not match the experimental data—there is only a modest (13.1%) requirement for increased regenerant—and even that modest increase is only required at the lowest regenerant dosage.

Softening systems that use more than 4 pounds of regenerant would therefore require no change to the amount of regenerant when converting from NaCl to KCl. Systems using 4 pounds of regenerant would need to increase the dosage from 4 pounds NaCl to 4.5 pounds of KCl when converting. Capacity of the softening system would then be unaffected by the change in regenerants.

COMPARISON OF KCl AND NaCl: USED AND UNUSED REGENERANT

In the preceding section we saw that the capacity differences between KCl and NaCl are very slight, and only significant at low regenerant dosages. We also observed that the capacity differences are much lower than expected based on theoretical (molecular weight) considerations. These results point to a more efficient use of regenerant when using

KCl for softening. In this section we examine the efficiency of the regenerants and compare the performance of KCl with NaCl.

We know that NaCl has the theoretical capacity of 5998 grains as $CaCO_3$ per pound of regenerant, while KCl has the theoretical capacity of 4702 grains as $CaCO_3$ (see the section "Theoretical Capacities of KCl and NaCl Regenerants").

Four pounds of regenerant would therefore have the capacity of:

$$4 \times 5998 \text{ grains} = 23,992 \text{ grains for NaCl regenerant}$$

$$4 \times 4702 \text{ grains} = 18,808 \text{ grains for KCl regenerant}$$

In contrast, we saw in Table 8-6 that 4 pounds of NaCl regenerant had a capacity for softening of 16,408 grains (as $CaCO_3$, per cubic foot of resin). The portion of the NaCl actually used for regeneration at 4 pound dosage is therefore:

$$\frac{16,408 \text{ grains}}{23,992 \text{ grains}} = 0.684 = 68.4\%$$

The portion of the NaCl unused during regeneration (and therefore flowing through the ion exchange resin, and reporting to waste) would therefore be the difference between the used portion and 100%, that is:

$$\text{Unused portion of NaCl} = 100 - 68.4\% = 31.6\%$$

Similarly, from Table 8-7 we saw that 4 pounds of KCl had an actual capacity for softening 14,505 grains (as $CaCO_3$, per cubic foot of resin). The portion of the KCl actually used for regeneration at 4 pound dosage is therefore:

$$\frac{14,505 \text{ grains}}{18,808 \text{ grains}} = 0.771 = 77.1\%$$

The unused portion of the KCl regenerant would then be:

$$\text{Unused portion of KCl} = 100 - 77.1\% = 22.9\%$$

Calculation of the portion of the regenerant actually being used for regeneration (and therefore reflected in the softening capacity) pro-

TABLE 8-9: Portion of KCl and NaCl used and unused during the regeneration cycle

Regenerant	Regeneration Pounds	Grains Exchanged per ft³, as CaCO₃	Percentage Used	Percentage Unused
NaCl	4	16,408	68.4	31.6
NaCl	8	25,585	53.3	46.7
NaCl	15	33,278	37.0	63.0
KCl	4	14,505	77.1	22.9
KCl	8	25,255	67.1	32.9
KCl	15	32,552	46.2	53.8

Data are calculated from Tables 8-6 and 8-7 as described in the text.

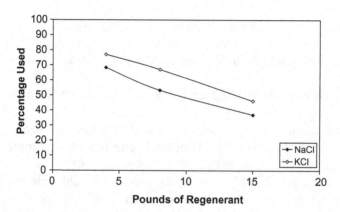

FIGURE 8-9 Percentage of regenerant used during regeneration, for both KCl and NaCl. Results are calculated from data provided in Tables 8-6 and 8-7 as presented in Table 8-9.

vides insight into the different capacities observed for NaCl and KCl. Theoretically, the capacity of KCl regenerant should be 27.6% less than that of NaCl. However, the above analysis shows that KCl is more effective as a regenerant, with 77.1% of the KCl actually being used during regeneration, compared to just 68.4% for the NaCl. Similar analysis for the other regenerant dosages given in Tables 8-6 and 8-7 was used to calculate the fractions used and unused; the results are provided in Table 8-9.

The percentage of the regenerant actually used during regeneration is plotted as a function of dosage in Figure 8-9 for both KCl and NaCl. The impact of efficiency is apparent from Figure 8-9; the higher-than-expected capacity of KCl as a regenerant is due to the greater percentage used during regeneration. The high percentage of KCl used suggests a higher affinity for K^+ adsorption onto the resin, relative to Na^+, which is consistent with the data provided in Table 4-1.

COMPARISON OF KCl AND NaCl: RELEASE OF CHLORIDES TO THE ENVIRONMENT

An interesting result of the study that compared NaCl and KCl as regenerants was the observed reduction of chlorides released to the environment when using KCl, which was not the anticipated outcome of the study. Review of the performance data, however, shows that the conclusion is valid.

Consider the data shown in Table 8-8. A regeneration cycle using 4 pounds of regenerant will have a capacity ratio of 1.131. To achieve the same softening capacity as 4 pounds of NaCl, one would have to use:

$$4 \text{ pounds of NaCl} \times 1.131 = 4.524 \text{ pounds of KCl}$$

Similar calculations can be done for 8 and 15 pounds of regenerant, so that we can tabulate the pounds of KCl required for softening equivalent to that of NaCl. The results are shown in Table 8-10.

Also tabulated in Table 8-10 is the amount of chloride released to the environment when regenerating with NaCl versus the chloride released when regenerating with the pounds of KCl required to give equivalent softening. All chloride present in the regenerant will eventually report to the environment within the spent regenerant solution. The amount of chloride released can therefore be calculated based on the known atomic and molecular weights.

The chloride present in 4 pounds of NaCl can be calculated since the amount of chloride will be proportional to the ratio of the atomic weight of chlorine to the molecular weight of NaCl (Chapter 5). Using

TABLE 8-10: Pounds of KCl required for softening equivalent to NaCl at regenerant levels of 4, 8, and 15 pounds. Also provided is the amount of chloride released to the environment, from these quantities of KCl and NaCl.

Pounds NaCl	Capacity Ratio	Pounds KCl	Chloride Released From NaCl (pounds)	Chloride Released From KCl (pounds)
4	1.131	4.524	2.427	2.151
8	1.013	8.104	4.853	3.854
15	1.022	15.33	9.099	7.290

data provided in Tables 5-1 and 5-2, we can determine the amount of chloride released in 4 pounds of NaCl:

$$\text{pounds chloride} = 4 \text{ pounds NaCl} \times \frac{\text{atomic weight chlorine}}{\text{molecular weight NaCl}}$$

$$\text{pounds chloride} = 4 \text{ pounds NaCl} \times \frac{35.453 \text{ g/mole}}{58.443 \text{ g/mole}}$$

$$\text{pounds chloride} = 2.427 \text{ pounds}$$

Similarly, we can calculate the pounds of chloride released from 4.524 pounds of KCl (which will provide softening equivalent to 4 pounds NaCl):

$$\text{pounds chloride} = 4.524 \text{ pounds KCl} \times \frac{\text{atomic weight chlorine}}{\text{molecular weight KCl}}$$

$$\text{pounds chloride} = 4.524 \text{ pounds KCl} \times \frac{35.453 \text{ g/mole}}{74.551 \text{ g/mole}}$$

$$\text{pounds chloride} = 2.151 \text{ pounds}$$

Thus, although the total weight of KCl is greater than NaCl for equivalent softening, the weight of chlorides released to the environment is less, because of the different ratios of atomic and molecular weights. The difference in chloride releases are even more pronounced for regeneration at the 8- and 15-pound levels. The complete set of data for chloride released to the environment is included in Table 8-10 and illustrated in Figure 8-10.

The data shown in Table 8-10 and illustrated in Figure 8-10 show clearly that the amount of chloride released to the environment is substantially (up to 20%) lower when KCl is used as a regenerant instead of NaCl for the equivalent amount of softening. In areas where sodium and chloride levels in water softener effluent are a concern, the use of potassium chloride as a regenerant is a very desirable alternative.

COMPARISON OF KCl AND NaCl: TASTE

Substitution of KCl for NaCl in home softening will result in the introduction of small, but significant, amounts of potassium in the home drinking water, along with a corresponding reduction in the amount of

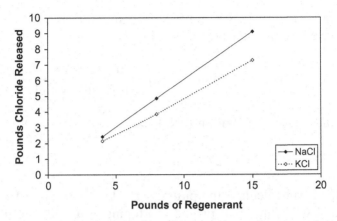

FIGURE 8-10   Total pounds of chloride released to the environment when regenerating with NaCl as a function of the regenerant dosage. For comparison, the chloride release is also plotted for regeneration with the amount of KCl required for equivalent softening.

sodium. People are very sensitive to the taste of their drinking water, so a small blind taste test was done to determine whether the substitution of potassium for sodium would be detected, and would be a cause for concern.

Standard drinking water samples were made from distilled water, and each sample was spiked with one type of salt. For each salt, eight different water samples were made, with the concentration of the salt in the samples varying from 100 to 800 parts per million (ppm) in 100-ppm increments. Six different salts were studied: KCl, $NaCl$, K_2SO_4, Na_2SO_4, $KHCO_3$, and $NaHCO_3$, which yielded a total of 48 drinking water samples.

Five subjects tasted each of the water samples in a blind taste test. The responses of the subjects were categorized as:

— No detectable taste
— Indeterminate flavor change detected
— Detection of salt flavor

The results of the taste test are shown in Tables 8-11 through 8-16. Included in these tables is the "taste index" for each water sample. The response of each subject was assigned a score of 0.0 for each "no detectable taste," 0.5 for each "indeterminate flavor change noted," and 1.0 for each "detection of salt flavor." The taste index was then calculated as the sum of the five scores for each water sample. The taste

TABLE 8-11: Response of subjects to a blind taste test for distilled water spiked with varying concentrations of potassium chloride (KCl)

Subject No.	Concentration of KCl in Distilled Water (ppm)							
	100	200	300	400	500	600	700	800
1	–	–	–	–	–	DC	DS	–
2	–	DS	–	–	–	–	–	–
3	–	–	–	–	DC	DS	–	–
4	–	–	–	–	–	–	DS	DS
5	–	–	–	–	DS	–	–	–
Taste Index	0	1.0	0	0	1.5	1.5	2.0	1.0

Responses are tabulated according to no detectable taste (–); indeterminate flavor change detected (DC), or detection of salt flavor (DS).

TABLE 8-12: Response of subjects to a blind taste test for distilled water spiked with varying concentrations of sodium chloride (NaCl)

Subject No.	Concentration of NaCl in Distilled Water (ppm)							
	100	200	300	400	500	600	700	800
1	–	–	DC	DS	DS	–	–	–
2	–	–	–	–	–	–	DC	–
3	–	–	–	–	–	DS	DS	–
4	–	–	–	DC	DS	–	–	DS
5	–	–	–	–	DC	–	–	DS
Taste Index	0	0	0.5	1.5	2.5	1.0	1.5	2.0

Responses are tabulated according to no detectable taste (–); indeterminate flavor change detected (DC), or detection of salt flavor (DS).

TABLE 8-13: Response of subjects to a blind taste test for distilled water spiked with varying concentrations of potassium sulfate (K$_2$SO$_4$)

Subject No.	Concentration of K$_2$SO$_4$ in Distilled Water (ppm)							
	100	200	300	400	500	600	700	800
1	–	–	–	DS	DS	DS	–	DS
2	–	–	–	–	–	–	–	DC
3	–	–	–	–	–	–	–	–
4	–	–	–	–	DC	–	DS	–
5	–	–	–	–	DC	DS	DS	–
Taste Index	0	0	0	1.0	2.0	2.0	2.0	1.5

Responses are tabulated according to no detectable taste (–); indeterminate flavor change detected (DC), or detection of salt flavor (DS).

TABLE 8-14: Response of subjects to a blind taste test for distilled water spiked with varying concentrations of sodium sulfate (Na_2SO_4)

Subject No.	Concentration of Na_2SO_4 in Distilled Water (ppm)							
	100	200	300	400	500	600	700	800
1	–	–	DC	DC	DS	DS	DS	DS
2	–	–	–	–	–	DC	–	DS
3	–	–	–	DC	–	–	–	DC
4	–	DC	–	DC	DC	–	–	–
5	–	–	DC	–	–	DS	DS	–
Taste Index	0	0.5	1.0	1.5	1.5	2.5	2.0	2.5

Responses are tabulated according to no detectable taste (–); indeterminate flavor change detected (DC), or detection of salt flavor (DS).

TABLE 8-15: Response of subjects to a blind taste test for distilled water spiked with varying concentrations of potassium bicarbonate ($KHCO_3$)

Subject No.	Concentration of $KHCO_3$ in Distilled Water (ppm)							
	100	200	300	400	500	600	700	800
1	–	–	–	–	DS	DS	DS	DS
2	–	–	–	–	DC	DC	DC	DC
3	–	–	–	–	–	–	–	–
4	–	DC	–	–	DS	DC	–	–
5	–	DC	DC	DC	–	–	DC	–
Taste Index	0	1.0	0.5	0.5	2.5	2.0	2.0	1.5

Responses are tabulated according to no detectable taste (–); indeterminate flavor change detected (DC), or detection of salt flavor (DS).

TABLE 8-16: Response of subjects to a blind taste test for distilled water spiked with varying concentrations of sodium bicarbonate ($NaHCO_3$)

Subject No.	Concentration of $NaHCO_3$ in Distilled Water (ppm)							
	100	200	300	400	500	600	700	800
1	–	–	DC	DS	DS	DS	DS	DS
2	–	–	DS	–	–	–	–	DC
3	–	–	–	DC	–	–	–	–
4	–	DC	–	DC	DS	–	DS	DS
5	–	DC	–	–	DC	–	–	–
Taste Index	0	1.0	1.5	2.0	2.5	1.0	2.0	2.5

Responses are tabulated according to no detectable taste (–); indeterminate flavor change detected (DC), or detection of salt flavor (DS).

index was determined to provide a simple measure of the overall flavor of the sample.

The data shown in Tables 8-11 through 8-16 suggest that low levels of KCl in drinking water do not have a major impact on taste. This point is illustrated in Figures 8-11 and 8-12. In Figure 8-11 the taste index of KCl is plotted along with five other inorganic salts, including NaCl. The perception of taste in water appears to follow a broad band, with no taste below approximately 200 ppm and the impact increasing in a general way with concentration. Although not a rigorous test, the data shown in Figure 8-11 suggest that both KCl and NaCl have less impact on taste than the other inorganic salts. In Figure 8-12 the taste impact of NaCl versus KCl is clarified by leaving out most of the other salts. There does not appear to be a major difference in the taste of

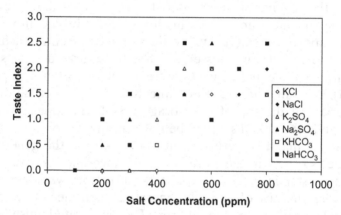

FIGURE 8-11 Taste index as a function of salt concentration for several different salts dissolved in distilled water.

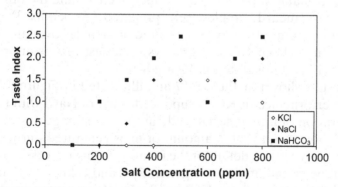

FIGURE 8-12 Taste index as a function of salt concentration for solutions of KCl, NaCl, or NaHCO$_3$ dissolved in distilled water.

water containing NaCl versus KCl. The data for sodium bicarbonate—the salt with the greatest taste impact—have been included in Figure 8-12 for comparison.

The comparison of taste between low levels of NaCl and KCl described in this section was based on a small sample of people. In addition, taste can be a subjective matter and difficult to quantify. Nonetheless, the results suggest that softening water with KCl instead of NaCl is not likely to have a major impact on the taste of the water for most people.

COMPARISON OF KCl AND NaCl: GENERATION OF FINES

One of the practical problems that can occur in household softeners is caking of the regenerant in the salt storage vessel. Caking can cause many problems including the formation of a salt bridge, in which the salt cakes above the water level in the storage vessel. In such a case, the water in the storage vessel will not be saturated with salt and regeneration will be ineffective (and treated water will not be properly softened). The salt bridge can be difficult to detect, because often the salt observed at the top of the storage vessel will appear to be free flowing, while the caked salt will be lower in the vessel.

One of the major underlying causes of caking is the generation of fine particles of salt in the storage vessel. Since fines generation is so important, a standardized "accelerated mush test" was developed to measure the quantity of fine salt generated when the softening salt is stored, and stirred, in a saturated brine. The accelerated mush test used in the present study is provided in Appendix 3.

The standard accelerated mush test was used to compare two commercially available water softening salts—one KCl and the other NaCl. A series of standard samples were prepared (Appendix 3), and the quantity of generated fines was measured on samples over an extended period of time (from 1 day to 28 days). The test results are provided in Table 8-17 and illustrated in Figure 8-13.

The results shown in Table 8-17 and illustrated in Figure 8-13 show that the performance of KCl is superior to that of NaCl with regard to the generation of fine material while in the salt storage vessel. It must be noted, however, that the amount of fines generated during storage will be sensitive to the details of the manufacturing process. As a result, there will be variability in fines generation, and caking, from one supplier to the next—for both KCl and NaCl. It is prudent for the consumer to ensure that their brand of regenerant is provided by a supplier

TABLE 8-17: Volume of fine salt (in cubic centimeters) generated during the accelerated mush test, for KCl and NaCl softening salt stored for various lengths of time in a saturated brine

Time (days)	Volume of Fine KCl (cc)	Volume of Fine NaCl (cc)
1	2.1	5.4
2	3.1	7.8
4	4.2	23.1
7	6.1	25.0
14	10.9	19.5
21	11.7	19.6
28	13.5	19.8

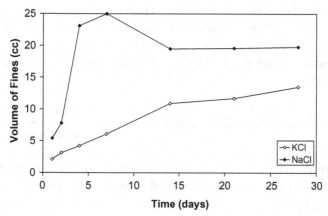

FIGURE 8-13 Volume of fine salt (in cubic centimeters) generated during the accelerated mush test for KCl and NaCl softening salt stored in a saturated brine, as a function of time.

who has adequate quality control practices to ensure that caking will not be a problem.

COMPARISON OF KCl AND NaCl: SODIUM CONTENT OF SOFTENED WATER

Softened water is often used for drinking, and the adverse health effects of excessive sodium are well known, so the sodium content of softened water is an important issue. In this section we therefore examine the sodium content of softened water, and compare the performance of sodium chloride and potassium chloride regenerants.

Consider an ion exchange resin in a water softener, freshly regenerated with NaCl. We know that the majority of the exchange sites on the resin will be loaded with sodium. Thus if we examine the ion exchange reaction (Chapter 3) we will see that the equilibrium is shifted to the left:

$$2R\text{--}SO_3^-Na^+ + Ca^{2+} \rightleftharpoons (R\text{--}SO_3^-)_2\, Ca^{2+} + 2Na^+$$

If the softener is then put into service, and hard water flows through the column, calcium and magnesium ions will adsorb onto the ion exchange resin and sodium ions will be released. As the above reaction shifts to the right, sodium ions are released into the effluent, and the concentration of sodium in the effluent thus rises.

From the above reaction we see that each hardness ion (calcium or magnesium) adsorbed will release two sodium ions. On a mass basis, we can then calculate the amount of sodium in the effluent due to the adsorption of calcium ions by using the atomic weights (Table 5-1):

$$\text{Sodium in effluent} = 2 \times \text{calcium in influent} \times \frac{\text{atomic weight Na}}{\text{atomic weight Ca}}$$

As an example, if a water sample has 241 mg/L of calcium, then adsorption of calcium ions by the resin will increase the sodium content of the effluent by:

$$\text{Sodium in effluent} = 241\,\frac{\text{mg}}{\text{litre}} \times 2 \times \frac{22.990}{40.08} = 276\,\frac{\text{mg}}{\text{litre}}$$

Adsorption of hardness ions is thus one major factor contributing to the sodium content of water softened with NaCl. Two other factors that will also contribute to the sodium content of the effluent are:

— Sodium in the influent. If the softener was regenerated with NaCl, then the influent sodium ions cannot initially be adsorbed by the ion exchange resin, since the resin is initially already fully loaded with sodium. As a result, the sodium ions in the influent will simply report directly to the effluent.

— Potassium in the influent. We know (Table 4-1) that potassium ions have a somewhat greater affinity for ion exchange resins than sodium ions. Therefore, when potassium ions in the influent pass through a column initially charged with NaCl, there will be an exchange of K^+ onto the resin and release of Na^+ into the effluent, according to the reaction:

$$R-SO_3^-Na^+ + K^+ \rightleftharpoons R-SO_3^-K^+ + Na^+$$

Unlike the softening reaction, the exchange of sodium for potassium is a simple 1:1 reaction. Thus the contribution to sodium in the effluent due to potassium will be a simple ratio of the atomic weights. Referring again to Table 5-1, we can calculate the contribution that potassium would make to the sodium content of softener effluent. If the influent water had a potassium content of 28 mg/liter, then the contribution to sodium in the effluent would be:

$$\text{Sodium in effluent} = 1 \times \text{potassium in influent} \times \frac{\text{atomic weight Na}}{\text{atomic weight K}}$$

$$\text{Sodium in effluent} = 28\,\frac{mg}{liter} \times 1 \times \frac{22.990\,g/mol}{39.098\,g/mol} = 16.5\,\frac{mg}{liter}$$

The concentration of sodium in softened water effluent will be the sum of the three contributing factors—sodium, potassium, and hardness ions in the influent. As an illustration, consider softening of a water sample with the composition:

Influent hardness $= H_{IN} = 35.2$ grains/gallon as $CaCO_3$

Influent sodium $= Na_{IN} = 66$ mg/liter

Influent potassium $= K_{IN}$ 28 mg/liter

We know that the initial sodium content of the effluent will be the sum of the sodium content from each of the three sources listed above. First, we use Table 3-2 to convert the hardness from grains per gallon to mg/liter:

$$H_{IN} = 35.2\,\frac{grains}{gallon} \times 17.1\,\frac{mg/liter}{grains/gallon} = 602\,mg/liter\ as\ CaCO_3$$

and then using the atomic and molecular weights provided in Tables 5-1 and 5-2 we can convert from $CaCO_3$ basis to calcium:

$$H_{IN} = 602\,\frac{mg}{liter}\ as\ CaCO_3 \times \frac{\text{atomic weight of calcium}}{\text{molecular weight of calcium carbonate}}$$

$$H_{IN} = 602 \frac{mg}{liter} \times \frac{40.08 \text{ g/mol}}{100.088 \text{ g/mol}} = 241 \frac{mg}{liter} \text{ as Ca}$$

We can then calculate the initial concentration of sodium in the softener effluent (Na_{EFF}) as the sum of the contributions from each of the three sources:

$$Na_{EFF} = Na_{IN} + \left(H_{IN} \times 2 \times \frac{\text{atomic weight Na}}{\text{atomic weight Ca}} \right) +$$
$$\left(K_{IN} \times 1 \times \frac{\text{atomic weight Na}}{\text{atomic weight K}} \right)$$

Substituting in the values of Na_{IN}, H_{IN} and K_{IN} given above, along with the appropriate atomic weights, we obtain:

$$Na_{EFF} = 66 \frac{mg}{liter} + \left(241 \frac{mg}{liter} \times 2 \times \frac{22.990 \text{ g/mol}}{40.08 \text{ g/mol}} \right) +$$
$$\left(28 \frac{mg}{liter} \times 1 \times \frac{22.990 \text{ g/mol}}{39.098 \text{ g/mol}} \right)$$
$$Na_{EFF} = 66 \frac{mg}{liter} + \left(277 \frac{mg}{liter} \right) + \left(16 \frac{mg}{liter} \right)$$
$$Na_{EFF} = 359 \frac{mg}{liter}$$

The initial sodium content of the effluent can be calculated with some accuracy, as shown above. As the softening process continues, the sodium content of the effluent will evolve because the resin will begin to become loaded with potassium and hardness ions. A significant amount of potassium will begin to flow through the softener into the effluent; as a result, there will be a concomitant reduction in the amount of sodium in the effluent. The selectivity of cation exchange resins is far higher for calcium and magnesium than for potassium, so it will be much longer before hardness ions occur in the effluent. Significant hardness in the effluent will only occur as the resin reaches capacity. Nonetheless, as hardness shows up in the effluent, there will be a concomitant reduction in sodium content. In the limiting case, as the capacity of the ion exchange system is surpassed, the sodium content of the effluent will return to the level of the influent—such an extreme situation would not occur in home softeners (the

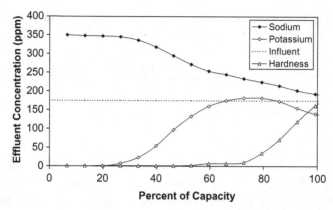

FIGURE 8-14 Composition of effluent from a water softener initially regenerated with sodium chloride as a function of the volume of water treated (expressed as a percentage of full capacity). Influent composition is described in the text.

system would be regenerated long before this time), but it illustrates the evolution of the softening system.

The evolution of sodium, potassium, and hardness ions in a typical softener effluent is illustrated in Figure 8-14 for an ion exchange column regenerated initially with NaCl, and which then processed an influent with the composition:

$$Na^+ = 175\,ppm$$

$$K^+ = 117\,ppm$$

$$Hardness = 230\,ppm\ as\ CaCO_3$$

Figure 8-14 is a good illustration of the performance of ion exchange columns, and it deserves some careful analysis. It can be seen that the initial effluent sodium composition is very high, since the ion exchange resin is adsorbing all of the hardness and potassium ions from the influent and replacing them with sodium. As shown above, we can calculate the initial sodium content of the effluent.

First, we note that 230 ppm hardness as calcium carbonate is equivalent to 92.1 ppm hardness as calcium. Then we can apply the formula for determining initial effluent sodium:

$$Na_{EFF} = Na_{IN} + \left(H_{IN} \times 2 \times \frac{atomic\ weight\ Na}{atomic\ weight\ Ca}\right) + \left(K_{IN} \times 1 \times \frac{atomic\ weight\ Na}{atomic\ weight\ K}\right)$$

Substituting into this equation the experimental values for influent composition, along with the known atomic weights, we obtain:

$$Na_{EFF} = 175\frac{mg}{liter} + \left(92.1\frac{mg}{liter} \times 2 \times \frac{22.990}{40.08}\right) + \left(117\frac{mg}{liter} \times 1 \times \frac{22.990}{39.098}\right)$$

$$Na_{EFF} = 349.5\frac{mg}{liter}$$

The calculated value of 349.5 ppm is in good agreement with the data illustrated in Figure 8-14 for the initial sodium content of the effluent. During this initial phase of the softening process, all hardness and potassium ions are adsorbed onto the resin, displacing sodium ions. The driving force for this reaction is the fact that calcium, magnesium, and potassium all have a higher affinity for the resin than sodium (Table 4-1).

Of the three ions, Ca^{2+}, Mg^{2+} and K^+, the potassium ions have the lowest affinity for the ion exchange resin (Table 4-1), and therefore as the softening process continues potassium is the first ion (other than sodium) to report to the effluent (Fig. 8-14). As potassium begins to report to the effluent, the amount of sodium in the effluent decreases, because the reaction:

$$R-SO_3^-Na^+ + K^+ \rightleftharpoons R-SO_3^-K^+ + Na^+$$

is no longer proceeding completely to the right.

As the softening reaction proceeds further, the amount of hardness in the effluent begins to increase dramatically at around 70% of capacity. Since fewer hardness ions are being adsorbed from solution, fewer sodium ions are displaced from the resin—the concentration of sodium in the effluent continues to fall, and starts to approach the concentration in the influent. Another interesting phenomenon is observed around 70% of capacity; at this point the concentration of potassium in the effluent peaks, and then begins to fall (eventually, of course, it would return to the same level as the influent). This strange behavior in potassium concentration is due to the relative affinity of the ion exchange resin for the ions: hardness > potassium > sodium. Initially, when the resin is fully loaded with sodium, all the potassium ions from the influent are adsorbed onto the resin and the concentration of potassium in the effluent is essentially zero. However, as softening proceeds further, and the resin begins to get loaded up with hardness ions, the potassium ions are displaced from the resin (by calcium and magnesium). Potassium ions loaded onto the column in the early stages of softening are stripped from the resin and move to the effluent; at this point the potassium content of the effluent exceeds that of the influent. By the time softening has reached 70% of capacity, most of the potassium has been stripped from the column, and the level in the effluent begins to approach that of the influent.

Thus far in this section we have had a close look at the behavior of loading, and stripping, of ions from an ion exchange resin loaded initially with sodium ions. If the ion exchange resin had been loaded initially with potassium, then we would expect similar behavior, except that the role of sodium and potassium would be reversed. This is, in fact, the case. An experiment was performed in which an ion exchange column was regenerated with KCl, and then used to process a hard water of composition:

$$Na^+ = 175\ ppm$$

$$K^+ = 117\ ppm$$

$$Hardness = 230\ ppm\ as\ CaCO_3$$

The composition of the effluent was monitored over the course of the softening process, and the results are shown in Figure 8-15.

Comparison of Figures 8-14 and 8-15 shows that, as expected, the behavior of the ion exchange column is very similar regardless of whether the resin is regenerated with KCl or NaCl; the only major

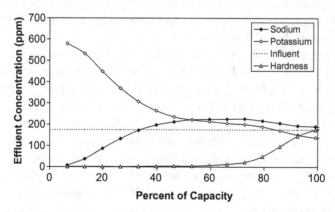

FIGURE 8-15 Composition of effluent from a water softener initially regenerated with potassium chloride as a function of the volume of water softened (expressed as a percentage of full capacity). Influent composition is described in the text.

difference is that the roles of K^+ and Na^+ ions are reversed when the column was regenerated with potassium chloride.

The close look at the behavior of loading and unloading of ions onto an ion exchange resin has now put us in a position to be able to understand the sodium content of the effluent—which is the goal of this section. For an ion exchange system initially regenerated with NaCl, we expect the initial sodium content of the system to be:

$$Na_{EFF} = Na_{IN} + \left(H_{IN} \times 2 \times \frac{\text{atomic weight Na}}{\text{atomic weight Ca}} \right) + \left(K_{IN} \times 1 \times \frac{\text{atomic weight Na}}{\text{atomic weight K}} \right)$$

with the sodium content of the effluent slowly tapering off to the same level as the influent when other ions—first potassium, followed by hardness ions—begin to report to the effluent.

In contrast, an ion exchange resin regenerated with KCl is expected to show very low sodium levels in the effluent initially as Na^+ is loaded onto the resin. As the softening process continues and the column starts to get loaded up with hardness ions, the Ca^{2+} and Mg^{2+} will displace sodium from the resin, so that a peak in sodium content (above the influent concentration) will be observed. Eventually, as the column becomes fully loaded, the composition of the effluent will of course approach the composition of the influent.

The expected sodium content of the softener effluent is, in fact, observed. The sodium concentration for the effluents illustrated in Figures 8-14 and 8-15 are plotted together in Figure 8-16. In this illustration we compare the sodium content of softener effluent for systems regenerated with NaCl versus KCl.

We can see from Figure 8-16 that the sodium content of softened water will be substantially lower for a system regenerated with KCl than with NaCl. In fact, if we calculate the average sodium content of the data points shown in Figure 8-16 we find that the system regenerated with NaCl had an average effluent content of 273 ppm Na^+, while the system regenerated with KCl had an average content of 171 ppm Na^+. In practice, a household softening system would be regenerated long before the softener approached 100% capacity, so the difference in sodium content (between KCl and NaCl regeneration) would be even more pronounced.

The impact of regeneration with KCl instead of NaCl is further illustrated in Figures 8-17 and 8-18. In Figure 8-17 the sodium content

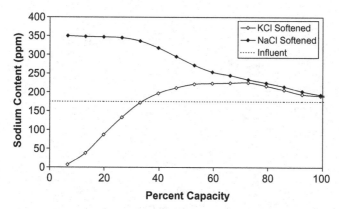

FIGURE 8-16   Sodium content of softener effluent as a function of the volume of water treated (expressed as a percentage of full capacity). The sodium content is compared for systems regenerated with NaCl and with KCl. For comparison, the sodium content of the influent water is also shown.

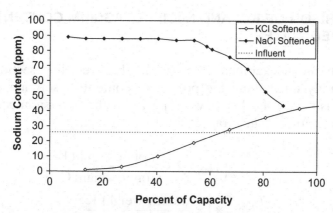

FIGURE 8-17   Sodium content of softener effluent as a function of the volume of water treated (expressed as a percentage of full capacity) for a system with low influent total dissolved solids (TDS).

of the softened water is examined for a system with a low influent TDS (26 ppm Na, 2 ppm K, 117 ppm hardness as $CaCO_3$). Figure 8-18 illustrates the impact of regenerant dosage on sodium content, and shows that the type of regenerant (KCl versus NaCl) is important, while the regenerant dosage is not.

The overall conclusion of this section is therefore that regeneration with KCl instead of NaCl will result in a substantially lower sodium content (and higher potassium content) of the softened water.

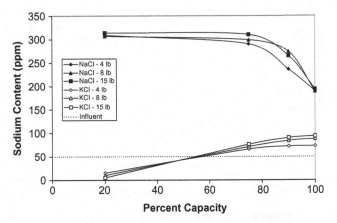

FIGURE 8-18 Sodium content of softener effluent as a function of the volume of water treated (expressed as a percentage of full capacity) showing the impact of regenerant dosage.

COMPARISON OF KCl AND NaCl: POTASSIUM CONTENT OF SOFTENED WATER

The phenomena examined in detail in the preceding section allow us to readily understand the potassium content of softener effluent. Analogous to the NaCl case, we expect that the potassium content of a softener initially regenerated with KCl will be:

$$K_{EFF} = K_{IN} + \left(H_{IN} \times 2 \times \frac{\text{atomic weight K}}{\text{atomic weight Ca}} \right) +$$
$$\left(Na_{IN} \times 1 \times \frac{\text{atomic weight K}}{\text{atomic weight Na}} \right)$$

and that the potassium content of the effluent will taper off to the influent level as sodium, and later hardness, ions begin to show up in the effluent. The expected effluent composition is, in fact, observed, as illustrated in Figure 8-19.

COMPARISON OF KCl AND NaCl: TOTAL DISSOLVED SOLIDS

One of the important parameters commonly used in water analysis is the total dissolved solids (TDS), which is the sum of the concentration of all salts dissolved in the water. Regeneration of an ion exchange

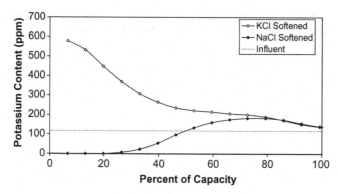

FIGURE 8-19   Potassium content of softener effluent as a function of the volume of water treated (expressed as a percentage of full capacity) showing the impact of regeneration with KCl versus NaCl.

system with KCl instead of NaCl will reduce the sodium content of the softened water, will release less chloride to the environment, and will achieve softening capacity essentially the same as NaCl. The measured TDS of the softened water will, however, be reported as a higher number. This is simply due to the fact that potassium has a higher atomic weight than sodium (Table 5-1).

Consider, for example, the softening of water that initially contains 150 mg/liter of calcium (reported as calcium). We know that the amount of sodium released into the softened water will be in proportion to the atomic weights, multiplied by 2 since the reaction between sodium and calcium is 2:1 (Chapter 4). The concentration of sodium in the effluent will therefore be:

$$\text{Sodium} = 150\,\frac{\text{mg}}{\text{liter}} \times 2 \times \frac{\text{atomic weight Na}}{\text{atomic weight Ca}}$$

$$\text{Sodium} = 150\,\frac{\text{mg}}{\text{liter}} \times 2 \times \frac{22.990\,\text{g/mol}}{40.08\,\text{g/mol}} = 172\,\frac{\text{mg}}{\text{liter}}$$

In contrast, if the softener had been regenerated with KCl, the concentration of potassium in the effluent would be:

$$\text{Potassium} = 150\,\frac{\text{mg}}{\text{liter}} \times 2 \times \frac{\text{atomic weight K}}{\text{atomic weight Ca}}$$

$$\text{Potassium} = 150\,\frac{\text{mg}}{\text{liter}} \times 2 \times \frac{39.098\,\text{g/mol}}{40.08\,\text{g/mol}} = 293\,\frac{\text{mg}}{\text{liter}}$$

TABLE 8-18: Chemical analysis, including total dissolved solids (TDS) for a typical hard water influent, along with the expected effluent analysis when softened with NaCl or KCl

Ion	Influent (mg/liter)	NaCl Softened Effluent (mg/liter)	KCl Softened Effluent (mg/liter)
Ca^{2+}	82	0	0
Mg^{2+}	20	0	0
Na^+	10	132	10
K^+	2	2	224
SO_4^{2-}	50	50	50
HCO_3^-	302	302	302
Cl^-	8	8	8
TDS	474	494	594

The ratio of potassium:sodium effluent concentrations will be the same as the ratio of the atomic weights:

$$\frac{293\,\text{mg/liter}}{172\,\text{mg/liter}} = \frac{39.098\,\text{g/mol}}{22.990\,\text{g/mol}} = 1.70$$

The impact of softening a typical hard water with potassium instead of sodium is illustrated in Table 8-18. Calculation of the data shown in this table has assumed that softening is complete. Moreover, it has been assumed that K^+ in the influent simply passes through the column for the case with regeneration with NaCl (and conversely, that Na^+ passes through when regeneration is done with KCl).

The reported TDS is higher for KCl softened water simply because of the higher atomic weight; this difference would have no effect on water quality.

CHAPTER 9

ENVIRONMENTAL CONSIDERATIONS

In Chapters 7 and 8 we saw that KCl is an effective regenerant for water softeners, with performance matching or exceeding that of NaCl. One important reason for considering the change from NaCl to KCl regenerant is the environmental impact of the softening process. Sodium has no redeeming qualities in the discharged waste. Potassium, in contrast, is an essential plant nutrient, so there is potential for the wise application of KCl regenerant to have a positive impact on the environment, or at least a less harmful impact than NaCl. In assessing the effects of different water softening agents on the environment, we must consider the uses and environmental fate of both the softened water and the discharged regenerant brine. The discussion in the present text is intended to be an overview of a complex subject. For more detailed information the enthusiastic reader is referred to the comprehensive review on the subject by Grasso et al. (1992).

The use of softened water results in the disposal of both the household water and the waste regenerant water into municipal sewage systems or septic systems. In some cases, cities reclaim water for sale to secondary markets, such as irrigation water to agricultural interests.

Water Softening with Potassium Chloride: Process, Health, and Environmental Benefits, by William Wist, Jay H. Lehr, and Rod McEachern
Copyright © 2009 by John Wiley & Sons, Inc.

On a smaller scale, such as households, planned communities, country clubs, etc., there is a growing interest in recycling graywater and using it for turfgrass and landscape irrigation. The kinds and amounts of dissolved materials in the water essentially dictate if and how it can be reused.

Regardless of whether it is household or recycled, discharged water, and possibly some dissolved solids (salts), may eventually enter groundwater or surface water. In some instances, the quality of the groundwater or surface water is changed enough to affect subsequent uses for agriculture and/or drinking.

As we develop and improve systems and strategies for handling discharged water, it is very important to look at alternative water treatments that enhance these new approaches. Potassium chloride as a water softening agent offers several major environmental advantages over sodium chloride in these systems. It also has several advantages in conventional discharge systems.

Several cities across the U.S. are taking aim at the water softening industry for discharging high levels of sodium and chloride into sewage treatment facilities. These cities are finding it hard to discharge reclaimed water without degrading groundwater and/or streams and rivers beyond acceptable limits. Some cities want to sell their reclaimed water to agriculture for crop irrigation. Since some crops are sensitive to sodium and chloride, the reclaimed discharge water may not be suitable for resale. Sodium chloride-softened water and the regenerant water both contribute to problems for these cities.

Switching to potassium chloride will reduce the sodium and chloride levels in discharge water. First, potassium will replace sodium in the softener, so no added sodium enters the system. Second, potassium chloride has a lower percentage (by weight) of chloride, per molecule, as compared to sodium chloride. In most softeners, this would equate to a 12 to 20 percent reduction in the amount of chloride discharged during each cycle. Third, the water will have a fertilizer benefit because of the enhanced potassium content. All of these points can be significant for a city trying to meet discharge standards and/or resell their graywater.

We can only assume that, in the future, the number of affected cities will increase and sodium chloride discharge concerns will be magnified. Potassium chloride is an excellent way to address these problems. In the following sections we examine and compare the impact of NaCl and KCl regenerants on the environment in several key areas. Later in the chapter, research results on the appropriate use of KCl regenerants in graywater irrigation will be presented.

POTASSIUM VERSUS SODIUM: IMPACT ON SOIL

The negative impact of excess sodium on soil properties is well known, and as a result the sodium content of soil has been well studied. The sodium content of a soil is generally described in terms of the sodium adsorption ratio (SAR), which is defined:

$$SAR = \frac{[Na^+]}{\sqrt{\frac{1}{2}([Ca^{2+}] + [Mg^{2+}])}}$$

where $[Na^+]$, $[Ca^{2+}]$, and $[Mg^{2+}]$ are the concentrations of sodium, calcium, and magnesium respectively, expressed in terms of milli-equivalents per liter (meq/l) in the soil solution. To measure the SAR of a soil sample, the material is first extracted with water to leach out soluble ions. The aqueous solution is then analysed, typically by atomic absorption, to measure the concentration of sodium, calcium, and magnesium.

The SAR is important to soil scientists because it quantifies the impact of excess sodium. Moreover, by factoring in the calcium and magnesium content of the soil, the SAR recognizes that these ions play a role in moderating the effects of the sodium. An SAR in excess of 12–15 will lead to serious problems with the physical properties of the soil, and plants will have difficulties absorbing water (Munshower, 1994).

The sodium content of soils is also described in terms of the exchange-able sodium content of the soil. To a first approximation, the minerals in soil can be considered similar to a cation exchange resin. That is, the surface of many minerals will adsorb cations, and these cations can be exchanged with other ions in solution. We can therefore describe the adsorption of ions on the surface of soil minerals using the same math-ematical models as used in Chapter 4 to describe ion exchange resins. Referring to Eq. 11 in Chapter 4, we can state that the fraction of avail-able sites occupied by sodium ions will be N'_{Na} defined as:

$$N'_{Na} = \frac{|Z_{Na}|[N_{Na}]}{N_{TOT}}$$

where Z_{Na} is the ionic charge of sodium ions (+1), $[Na^+]$ is the concen-tration of sodium (moles/liter), and N_{TOT} is the total exchange capacity of the soil, in equivalents. In soil science terminology, the fraction of available exchange sites occupied by sodium ions, N'_{Na}, is known as the exchangeable sodium ratio (ESR) and when expressed as a percentage is called the exchangeable sodium percentage (ESP). Not surprisingly, the ESP of a soil shows a positive correlation with SAR, with the proportionality constant known as the Gapon coefficient (Bower, 1959). The relationship between ESP and SAR will vary from one soil type to another because of different soil mineralogy, that is, the Gapon coefficient will vary from one region to another.

In the United States, soils with an ESP of >15 are considered sodic (a soil containing levels of exchangeable sodium that will interfere with the growth of plants), while soils with an ESP >30 will be highly sodic and will display a serious impact on plant growth (Sparks, 2003). Soils with an accumulation of excess sodium generally display poor tilth and low permeability.

It is well known that excess sodium damages clay-rich soils by reduction of the hydraulic conductivity. Sodium adsorbs to the surface of clays, resulting in swelling of expansible clays and dispersion of non-expanding clays. Sodium disperses clays by modifying the surface charges so that the small clay particles disperse into solution, rather than adhering to aggregates. The small clay particles that have dispersed clog up pores in the soil and degrade soil structure. The result is a soil that restricts root growth and water movement (hydraulic conductivity).

In addition to high sodium levels, the hydraulic conductivity can be affected by low levels of electrolytes and high pH values. The relationships between SAR, electrolyte concentration, pH, and hydraulic conductivity of soil have been studied extensively and are generally accepted (Keren and Singer, 1988; Chiang et al., 1987; Pupisky and Shainberg, 1979). It must be noted, however, that the impact of soil parameters such as SAR on hydraulic conductivity will vary among different soils because of variations in clay mineralogy from one region to the next (Velasco-Molina et al., 1971) so conclusions need to be drawn carefully.

Thus far in this section we have had a close look at the impact of excess sodium on soil properties. Clearly, excessive sodium in the soil has adverse effects on soil physical properties (especially hydraulic conductivity) and crop yields. The most important issue for this text, however, is to determine whether excess potassium would have the same deleterious impact on soils. We shall examine this question in the following paragraphs.

The potassium content of soils is described in terms of the exchangeable potassium ratio (EPR), exchangeable potassium percentage (EPP), and potassium adsorption ratio (PAR), where these terms are defined similar to the sodium analogs described above. It is well established that soils have a higher affinity for potassium ions than for sodium ions (Levy, 1995), which is similar to the adsorption character-istics of ion exchange resins (Table 4-1). Thus, for example, if a given soil had a PAR of 1.0, one would expect that the EPP would be greater than the ESP if the same soil had an SAR of 1.0. The impact of adsorbed potassium on soil properties has been examined by many researchers over the past number of years. The results are somewhat conflicting, perhaps because of different clay mineralogy in the various soils examined.

The impact of K^+, Na^+, Ca^{2+}, and Mg^{2+} on the hydraulic conductivity of tropical red earths and tropical black earths was studied by Ahmed et al. (1969). They reported that potassium gave a structural deteriora-tion similar to sodium, and they ranked the ions in order of impact: $Na^+ \geq K^+ > Ca^{2+} = Mg^{2+}$. Similarly, Martin and Richards (1959) reported a reduction in hydraulic conductivity with increasing potassium content. In contrast, numerous authors have reported that the impact of excess potassium is less deleterious than sodium on soils. Cecconi et al. (1963) and Ravina (1973) reported that potassium-rich soils displayed less clay dispersion and larger, more stable aggregates than sodium-rich soils; both of these studies reported that exchangeable potassium would have a more favorable impact on soil permeability than calcium. Similarly, Reeve et al. (1954) and Brooks et al. (1956) reported that potassium appeared to favor soil structural stability.

Many studies have reported that potassium has a deleterious impact on soil properties, but that its impact is less pronounced than that of sodium. Gardner et al. (1959) reported that potassium caused only 25% as much reduction as sodium when soil was treated with equal amounts of each cation. Chen et al. (1983) found that addition of low levels of potassium (i.e., low EPP) caused an increase in hydraulic conductivity in some soils and a reduction in others. However, they also reported that high EPP values resulted in the degradation of soil properties in all cases. They found that the hydraulic conductivity of soil was reduced by 80% when the ESP was 15%, but, in contrast, a similar reduction in hydraulic conductivity was only found for potassium when the EPP was 60–70%.

One of the reasons why various studies draw different conclusions about the relative impact of sodium and potassium is that clay mineral-ogy varies from one soil to the next. In particular it appears that inter-actions of K^+ may be more favorable if the soil clays are smectic rather than kaolinitic or illitic. Hesterberg and Page (1990) found that potas-sium was three times more effective than sodium at coagulating illite because of the greater tendency for K^+ to adsorb onto the clay surface.

These results are therefore consistent with the tendency for potassium to be less dispersive on soils than sodium.

Several conclusions can be drawn from the above discussion comparing the impact of sodium and potassium on soil physical properties. Damage to soils from excess sodium or potassium will generally be indicated by low permeability and hydraulic conductivity. The deleterious impact of excess sodium on soils is well documented. In contrast, there is conflicting evidence on the impact of excess potassium, which suggests that the impact of potassium may well be dependent on the nature of the soil clay mineralogy. On the balance of the available scientific studies, it appears that substitution of potassium for sodium in regenerant wastewaters will have a neutral impact, or quite possibly a less deleterious impact on affected soils.

POTASSIUM VERSUS SODIUM: IMPACT ON SEPTIC SYSTEMS

Septic systems are used for private homes that are not connected to a municipal sewage system. Wastewater is transferred to a holding (septic) tank that has capacity to hold 3–5 days of wastewater. In the septic tank organic matter such as oils and greases forms a scum layer on the top of the fluid. Similarly, heavier materials sink to the bottom to form a sludge layer. Bacteria in the septic tank digest some of the wastes, resulting in a clarified and partially purified fluid that is then discharged to the septic drainage field.

When considering the impact of regenerant waste on septic systems (tank and field) one must consider both the bacterial growth in the septic tank as well as the long-term impact on soil properties in the septic field.

A healthy bacterial ecosystem is necessary in a septic system for solids settlement and proper digestion of wastes. Concerns were raised that the increased sodium levels due to water softening could have a harmful impact on the bacteria in septic systems. Studies were therefore done by the National Sanitation Foundation (NSF) and the University of Wisconsin to determine whether water softeners have a deleterious effect on septic systems (Moore, 2001). The studies showed that there were no adverse impacts on microorganisms exposed, and the use of water softeners in households with septic systems is widely considered a safe practice. A subsequent study by the Centre for Water Resources Studies (2001) supported this conclusion and added that increased salt levels may even encourage bacterial growth in the septic system because it reduces the osmotic potential difference between the wastewater and the bacteria. Other studies, however, indicate some concern that an increased sodium loading could be harmful to the bacterial population in a septic system (Moore, 2001; Grasso et al., 1992). Tyler et al. (1978) reported that addition of regenerant brine and backflush to the septic system did not have a deleterious impact on the bacteria in the septic system. However, they also reported that the increased sodium loading led to a lag time for gas production, and that the system required time for acclimation, both of which are indicative of impaired bacterial growth.

The bulk of scientific evidence thus suggests that water softeners (using either KCl or NaCl) do not pose a substantial threat to the viability of the bacteria in the septic tank. However, there remain some concerns, and it is possible that impairment of bacterial growth could occur because of excessive sodium loading. Much of the sodium burden in household wastewater originates from materials other than softener regenerant, so it seems likely that conversion of the softener from NaCl to KCl could only have a positive impact on the bacteria in the septic tank.

Water softener regenerant and backflush are often added to household septic systems, and potential harm to soils in the septic field has been the subject of a number of studies. Tyler et al. (1978) examined the impact of addition of softener regenerant into a septic system. They concluded that the regenerant waste, which is rich in calcium and magnesium chlorides, would help to balance the increased sodium (or potassium) loading caused by adding softened water to the septic system. Addition of the regenerant waste would therefore reduce the risk of an increased sodium adsorption ratio (SAR) and the resulting decrease in soil hydraulic conductivity (in the septic field) as discussed in the preceding section.

In contrast to the work by Tyler et al. (1978), several other authors studied the impact of adding regenerant waste to septic systems and reported that such a practice will eventually lead to groundwater contamination, damage to the septic field soil characteristics, or both (Siegrist, 1988; Robertson et al., 1991; Mancino and Pepper, 1992). In these studies groundwater contamination by sodium and/or potassium was identified, which suggested that increased Na^+ or K^+ loading to the system would be harmful. DeWalle and Schaff (1980) reported increased levels of calcium in the groundwater, which was attributed to sodium from the septic system adsorbing onto the soil minerals, thereby releasing calcium to the groundwater.

The above discussion shows that there is not unanimous agreement in the scientific community on the merits of regenerant discharge into a septic system. Over years of service, the use of home softeners (with or without addition of regenerant to the system) clearly leaves open the potential for soil degradation and groundwater contamination. There does not appear to be a direct scientific comparison between KCl and NaCl regenerant with regard to the impact on septic systems. The use of KCl regenerant should therefore be preferred from an environmental perspective for the following reasons:

— The impact of excess K^+ on soils is equivalent, or possibly less deleterious than NaCl, as discussed in the preceding section.

— Accumulation of excess K^+ over long periods of time is less likely than Na^+ because some of the potassium will be taken up by lawn and garden plants, and thereby removed from the system.

— Potassium ions in the septic field will be more readily adsorbed by soil minerals (see preceding section), resulting in more localized contamination, and less likelihood of groundwater contamination.

— Sodium salts are far more common in household use than potassium. For example, a recent study found that 38% of the sodium burden in septic systems was from laundry detergents. Therefore, it seems clear that a reduction in the sodium loading to the septic system and field by converting the softener to KCl regenerant will have a positive impact on the bacteria in the septic tank, as well as the soil properties in the septic field.

POTASSIUM VERSUS SODIUM: IMPACT ON SEWAGE TREATMENT SYSTEMS

The vast majority of sewage in North America is treated in large publicly owned treatment facilities. Much of the waste regenerant from water softeners eventually makes its way into such facilities, so the comparison of KCl and NaCl in such facilities is important.

Primary sewage treatment is largely a physical process, and so the impact of either KCl or NaCl is expected to be slight. Both K^+ and Na^+ have the ability to improve coagulation of suspended solids somewhat, but the impact is slight compared to divalent or trivalent ions. Otherwise, the impact of either KCl or NaCl will be small, and it is not likely that substitution of NaCl with KCl would have any impact.

The impact of KCl and NaCl on secondary, that is, biological sewage treatment has been studied extensively. The main purpose of secondary treatment is to reduce the biological oxygen demand of the wastewater through microbial digestion, so that discharge of the effluent to the environment will not have a significant impact on downstream waterways. The comparison of KCl and NaCl in secondary treatment therefore focuses on the impact of these salts on microbial activity.

The impact on secondary sewage treatment of NaCl, and comparison to KCl, has been widely studied (Kugelman and McCarty, 1965; Kincannon and Gaudy, 1966; Ludzack and Noran, 1965; McCarty, 1964). Kugelman and McCarty (1965) studied the impact of NaCl and KCl on anaerobic digestion over the ranges 0.046–9.2 g/L and 0.195–14 g/L, respectively. They found that each ion had an optimum range for anaerobic digestion. For Na^+ the optimum was 0.092–0.207 g/L, while for potassium it was 0.195–0.390 g/L. They also reported similar concentrations for initial growth inhibition: 3.45 g/L for sodium and 2.34 g/L for potassium. The work of Kugelman and McCarty (1965) thus suggests very similar impact between KCl and NaCl on anaerobic microbial activity, although they did note that when optimal concentrations were exceeded the impact of potassium was greater than that of sodium.

Kincannon and Gaudy (1966) compared the impact of KCl and NaCl on the activated sludge process and reported that the efficiency of the microbial process was inversely proportional to salt concentration above 3.5 g/L for both KCl and NaCl. Ludzack and Noran (1965) also examined the impact of NaCl concentration on the activated sludge process. They reported improved performance as the concentration was increased up to 5.0 g/L, but reduced flocculation, biological oxygen demand reduction, and denitrification at higher concentrations. They concluded that the reduction in microbial growth at high NaCl concentrations was due to the chloride ion, but this claim differs from the results of McCarty (1964), who reported that the cations were responsible.

An interesting study by Imai and Endoh (1990) found that increased levels of potassium resulted in more efficient uptake of phosphorus in an activated sludge process. Moreover, Comeau et al. (1987) found similar results for potassium, but also reported that a similar response was not observed for sodium.

In recent years studies have focused on the impact of K^+ or Na^+ on specific microorganisms. Some interesting results have emerged, including the work done by Sprott et al. (1984, 1985). They examined methanogenic bacteria, which normally have an internal K^+ concentration of 7.8 g/L, and suggested that the bacteria would generally be impacted at higher concentrations of K^+ because of interference with cell membrane potentials. Interestingly, the results reported by Sprott et al. are reasonably close to the concentration (3.5 g/L) reported by Kincannon and Gaudy (1966) above which microbial inhibition occurred.

The studies discussed above suggest a general trend that both anaerobic and activated sludge secondary treatments have similar impacts and optimal concentrations for KCl and NaCl. There is some evidence that suggests excess potassium may be helpful in uptake of excess phosphate, while sodium has no influence; however, more study is required before firm conclusions can be made.

POTASSIUM VERSUS SODIUM: SEWAGE SLUDGE

Many large sewage facilities operated by municipalities and industrial facilities generate substantial quantities of sewage sludge each year. This sludge is often applied to agricultural land to recover nutrient values; such application of sewage sludge is known as land farming.

The practice of land farming sewage sludge has been performed for many years, and potential impacts on soil properties and bacteria have been examined in numerous studies (Henry et al., 1954; Lui, 1982; Overman and Leseman, 1982; Paredes et al., 1987). There is general agreement in the literature that sodium from sewage sludge is a potential contaminant that can have a negative impact on soil properties and adversely affect crop yields. Mechanisms for damage to soils from sodium in sewage sludge would be similar to those described in the preceding sections.

In contrast to sodium, potassium in sewage sludge was not identified as having a potential negative impact. In fact, the studies showed that potassium (along with nitrogen and phosphorus) was effectively taken up by the crops, which even resulted in a significant increase in crop yields (Henry et al., 1954; Sommers, 1977). As a result of the effective uptake of potassium and other nutrients by the crops, it is not surprising that sewage sludge was found to cause groundwater contamination for sodium, but not potassium (Overman and Leseman, 1982). The bulk of scientific evidence therefore clearly supports the view that conversion to KCl regenerant from NaCl would have a positive impact on land farms that utilize sewage sludge, because of effective uptake of potassium by crops.

POTASSIUM VERSUS SODIUM: ALGAE GROWTH

Algae is a term used to describe a large diverse group of simple autotrophic eukaryotes. They can be single-celled or multicellular organisms, with the largest being termed "seaweed." Algae are photosynthetic organisms, like land plants, but lack the complex structures such as phyllids and rhizoids in nonvascular plants or leaves, roots, and other organs found in tracheophytes.

Algae growth is commonly thought of as negative because of association with excessive algae growth (an "algae bloom") that can lead to eutrophication of the water body. However, normal algae growth is essential to a healthy water way. As photosynthetic organisms, algae are the primary energy supply, and they oxygenate the water. Microscopic animals (zooplankton) feed on the algae, and the zooplankton in turn support larger forms of life such as insects and fish.

Pond fertilization is sometimes done to promote healthy algae growth in ponds and lakes. Fertilization can be done with either organic or inorganic fertilizers (Conte, 2000). When fertilizing to encourage algae growth, phosphorus and nitrogen are the two key elements that must be increased. Potassium is added occasionally, but increased potassium levels are generally regarded as optional for good algae growth (Conte, 2000). Two studies on existing algae populations showed that increased algae growth occurred as the potassium concentrations in the water increased (Kappe and Kappe, 1971; Corbitt, 1990). Neither study clarified the metabolic role that potassium played in the increased algae growth. Kappe and Kappe (1971) also examined the impact of increased sodium level and reported that it, too, could potentially have a positive impact on algae growth.

Excess algae growth, commonly referred to as an algae bloom, occurs when there is a surplus of nutrients in the water body. The most common cause of algae blooms is excessive phosphorus, which can result from human, animal and industrial wastes, as well as runoff from crops and lawns. Excessive algae growth can lead to oxygen depletion because a thick layer of algae can form on the surface of the water, which limits the depth to which sunlight can penetrate. Photosynthetic organisms below the thick algae layer will be unable to generate oxygen, and may die. Decomposition of the organic matter accentuates the problem, because the decomposition process consumes additional oxygen. Oxygen depletion due to algae blooms will lead to the death of fish and other aquatic organisms, and will impart a foul taste and odor to the water. Excessive nutrient levels and algae growth along with oxygen depletion is termed "eutrophication" (Carstensen et al., 2007).

The prokaryotic cyanobacteria, commonly known as "blue-green algae," are not strictly algae (Nabors, 2004). Nonetheless, blue-green algae are included in the present discussion because "algae blooms" from these organisms are often particularly troublesome. Cyanobacteria are capable of nitrogen fixation, so after an algae bloom of these organisms decomposition of the organic matter replenishes the nitrogen content of the waterway, increasing the likelihood of subsequent algae blooms. Cyanobacteria algae blooms can lead to eutrophication of the waterway, causing harm to aquatic life, but in addition the water can cause skin and respiratory tract irritation to people who enter the water. Moreover, cyanobacteria produce toxins that can harm animals and people who drink the untreated water.

The impact of excess sodium and potassium on algae-induced eutrophication was studied by Mohleji and Verhoff (1980). They measured phosphorus uptake as an indicator of green algae growth, and reported that sodium (in the absence of potassium) appeared to increase growth. In contrast, potassium (in the absence of sodium) appeared to decrease growth. They suggested that the formation of a sodium-phosphorus complex, which facilitates transport of phosphorus across the cell membrane, may explain their observations. Seale et al. (1987) reported similar results for blue-green algae; they found that sodium concentrations above 5 mg/L stimulated increased uptake of nutrients, especially phosphorus. Subsequent studies have confirmed the relationship between sodium concentration and the growth of blue-green algae (Brownell and Nicholas, 1967; Ward and Wetzel, 1975).

Scientific evidence suggests that both potassium and sodium are essential nutrients required for normal algae growth. However, excessive algae growth leading to eutrophication of the water body correlates positively with sodium concentration, but not with potassium. Recent research goes further, and suggests that increased potassium concentrations in water may inhibit excessive algae growth (Shukla and Rai, 2006, 2007) and that potassium could be used to regulate algae blooms in an eco-friendly manner. Clearly the bulk of scientific evidence supports the view that substitution of NaCl with KCl regenerant in water softening could be helpful in reducing excessive algae growth.

Consumption of water softening salt is in excess of two million tons annually in North America, and the bulk of this material eventually ends up in our water sources. Algae blooms are a major environmental issue around the world, and the scientific evidence suggests that sodium is a contributing factor. Conversion of home softeners to KCl could have a substantial impact on reduction of algae blooms, through reduced sodium loading to the water ways as well as by inhibition of excess growth due to increased potassium levels in the water.

POTASSIUM VERSUS SODIUM: IMPACT ON PLANTS AND ANIMALS

Sodium and potassium play an important role in the biochemistry of plant and animal life. Organisms establish and regulate gradients of Na^+ and K^+ between the cytoplasm and the surrounding environment by cation pumps, and the effective regulation of sodium and potassium is critical to survival of the organism.

Many prokaryotes (simple organisms like bacteria whose DNA is not contained within a nucleus) can grow in the absence of sodium (Strobel and Russell, 1991) but marine bacteria cannot (Berthelet and MacLeod, 1989). The impact of NaCl and KCl on the growth of bacterial organisms has been studied extensively. Low levels (<2.3 mg/L) of sodium chloride are known to stimulate microbial activity, while higher concentrations reduce microbiological activity (Berthelet and MacLeod, 1989; Strobel and Russell, 1991; Yang and Drake, 1990; Page, 1991). Studies on the impact of potassium chloride have shown a similar impact (Markevics and Jacques, 1985; Kirk and Doelle, 1992; Onoda et al., 1992) at concentrations similar to sodium chloride. Substitution of NaCl regenerant with KCl regenerant is therefore not expected to have any broad impact on microbiological activity in the environment (with the exception of septic systems, as discussed above).

Low levels of sodium and potassium are ubiquitous in aquatic environments, so it is difficult to determine the impact of a modest increase in the concentration of Na^+ or K^+ because of the choice of a water softener regenerant. Nonetheless, several studies have examined the impact of sodium and potassium on aquatic life. A study of bioaccumulation in aquatic organisms showed that potassium could be a factor in enhanced uptake of heavy metal toxins (de March, 1988). Fisher et al. (1991) reported that potassium can be toxic to *Dreissena polymorpha* (zebra mussel) with an LC_{50} of 138 mg/L, while being nontoxic to other species such as snails, mosquito fish, and freshwater mussels. The efficacy of potassium for controlling zebra mussels was confirmed by Wildridge et al. (1998), although they reported a substantially higher LC_{50} of 400 mg/L when a 96-hour recovery period was applied. Nonetheless, the toxicity of K^+ to zebra mussels has found widespread application in industrial water systems (Lewis et al., 1997) and in eradication efforts (Dietz et al., 1995; Virginia Department of Game and Inland Fisheries, 2005).

Several studies have shown that potassium can have a positive impact on aquatic vegetation (Strauss, 1976; Olafsson, 1980; Volt and Phinney, 1968), while sodium does not. The relationship between vegetation and potassium content is not surprising, given that it is an essential nutrient. Volt and Phiney (1968) reported a minimum potassium concentration of 2.5 mg/L to support aquatic vegetation.

The most profound environmental difference between sodium and potassium is, not surprisingly, the effect these elements have on plants. The harmful effects of sodium on soil physical and chemical properties were discussed in the preceding section on soils. In addition to the harm that sodium causes to the soil, excess sodium can cause damage to leaves and defoliation (Rhoades, 1972). The harmful impact of sodium on irrigation water has been well documented in the literature, and it is well known that high sodium content results in low crop yields and quality (Bower et al., 1968; Bajwa and Josan, 1989; Rhoades, 1972). In contrast to sodium, potassium is an essential nutrient for plants. Uptake occurs by diffusion of soluble potassium to the plant roots, where there is efficient and rapid uptake. The amount of potassium in soil moisture is quite low, and therefore potassium is regularly replenished in fields by fertilizer application of KCl (potash). Excess potassium in crops is not common, because of its efficient uptake, but excessive application can lead to a nutrient unbalance, which results in lower crop yields and quality (Palazzo and Jenkins, 1979; MacLean, 1977; Kresge and Younts, 1962). However, the comparison between sodium and potassium is striking when considering effects on plants: sodium has a toxic effect that reduces yields and quality, while potassium is an essential nutrient.

POTASSIUM VERSUS SODIUM: USE OF RECYCLED GRAYWATER

Untreated household wastewater that has not come into contact with toilet waste is defined as graywater. It includes water from bathtubs, showers, wash basins, and washing machines. It does not include waste water from kitchen sinks and dishwashers. Graywater may or may not be treated to remove odors and kill disease-causing organisms, depending on the complexity of the system and the intended use of the recycled water.

The simplest systems collect wastewater in a holding tank and then pump it onto lawns and landscape plants. The amount of sodium and chloride in the graywater is a significant factor in determining whether it can be used on certain plants and how much it might have to be diluted with clear water to be safe to plants. Substituting sodium chloride with potassium chloride will, again, essentially eliminate the sodium and significantly reduce the chloride. If dilution is necessary, it will be less with potassium chloride.

Potassium and sodium have different effects on soils and plants, particularly when they are present in excessive amounts. First, potassium is a critical nutrient for plant growth. Potassium is necessary for a large variety of chemical reactions and water relations in plants. Plants take up and utilize large amounts of potassium from soils as they grow. Potassium that is not taken up by plants is held in the soil by clay and organic matter and is not likely to percolate into groundwater. If potassium-softened water is used for house plants, it is advisable to apply volumes large enough that allow pots to drain, to prevent potassium buildup. Two to three waterings a year with potassium-softened water would supply enough potassium to meet plant potassium needs.

Sodium is not an essential plant nutrient and will interfere with plant growth when taken up in excessive amounts. Also, high levels of sodium will cause swelling of the clays in soil. This swelling reduces pore space and results in poor soil structure, poor soil aeration, and reduced water movement. Soils high in sodium (sodic soils) are of marginal value for crop production and are difficult to manage for turfgrass and ornamental plants as well. Water softened with sodium chloride should not be used to water house plants since the buildup of sodium in the soil can inhibit plant growth.

RECYCLING REGENERANT WASTEWATER

In home regenerating softening systems the regenerant water passing through the resin tank is very high in total dissolved solids. This solution contains high levels of sodium and chloride from the brine as well as high levels of calcium and magnesium coming off the resin. Cities that want to reuse graywater or cannot meet discharge standards are proposing bans on home regenerating water softeners because this is an easily identified source of salts, and they recognize that regenerant water is high in dissolved solids (salts).

A couple of approaches to the disposal of regeneration waste water are being proposed. One is to collect regeneration water in a third home tank and then transfer this to tanker trucks for disposal at remote sites. If potassium chloride were used as the conditioning salt, this water could be used as potassium fertilizer material. The fertilizer value would be approximately 40 to 50 pounds of potassium per 1000 gallons of regenerant water. Appropriate application rates of this material must be determined by a professional agronomist, because the existing potassium soil test levels and crops grown would determine its use.

Another approach is to collect the regenerate waste water in a third tank, dilute it with nonsoftened water and apply it to well-established lawns. There are several important issues that must be investigated for such a practice to be effective. The acceptable dilution ratio for the regenerant solution must be determined, as well as the number of waterings (with nontreated water) required between applications of the regenerant solution. In addition, the suitability of various plant species to the use of dilute regenerant waste must be determined. Research has been performed to address many of these issues, and the results are presented in the following sections. As the practice of using KCl regenerant waste for irrigation becomes more popular, further research will no doubt be done.

USE OF REGENERANT WASTEWATER STUDIES AT UNIVERSITY OF CALIFORNIA, DAVIS

A proposed solution to the disposal of KCl regenerant wastewater in home softening systems is to collect the wastewater in a third tank, dilute it with fresh water and apply it to well established lawn, garden, and/or ornamental plantings. There are several critical questions in this type of system. How much should the wastewater be diluted? Are there certain grasses or plants that are overly sensitive to this water? How much nontreated water should be used for irrigating between wastewater applications? What are the long-term effects of this approach on the soil and vegetation?

A multiyear research project with Dr. Lin Wu at the University of California, Davis was performed to answer many of these questions. The heart of the project was the use of potassium chloride rather than sodium chloride as the regenerant salt. With KCl, the regenerant waste water contains high levels of Calcium (Ca^{2+}), magnesium (Mg^{2+}), chloride (Cl^-), and potassium (K^+). The system is free of sodium, and the chloride levels are lower in the KCl system than in NaCl systems, as discussed in Chapter 8.

The objective of Dr. Wu's research was to determine the effects of high concentrations of Cl^-, Mg^{2+}, Ca^{2+}, and K^+ on the growth of turfgrass and ornamental species. He also studied the amounts of nutrients being taken up and removed in grass clippings, and the effects of these nutrients on the soil itself. The research project generated valuable information for the safe use and disposal of home softener wastewater. The results of his studies are presented in the following two sections for turfgrasss and ornamental plant irrigation.

Turfgrass Irrigation

Two warm season turfgrass species (Bermuda grass and Zoysia grass) and three cool-season species (tall fescue, Kentucky bluegrass, and perennial rye grass) were tested in the University of California, Davis, study. The study evaluated the impact of irrigation with wastewater produced by a softener regenerated with KCl at various dilution ratios. The impact of irrigation with regenerant wastewater was compared to irrigation with tap water. Results of the study have been published in the article "Regenerant wastewater irrigation and ion uptake by five turfgrass species" by L. Wu et al. (1996).

Field plots were represented by 16-inch diameter and 18-inch deep plastic containers filled with clay-loam soil. The containers were buried in a ten foot by forty-five foot concrete slab up to the height of the containers. Separate drainage was installed for each container so drain water could be collected and all added nutrients could be accounted for. In the first week of August, 1993, the field plots were sodded with the five turfgrass species, with three replications of each. The plots were fertilized with nitrogen and irrigated with city water for 8 weeks, to get them well established. The turf was mowed to a 2-inch height weekly. Throughout the test period, the established turfgrass was fertilized with a complete granular fertilizer every 2 months.

Wastewater treatments began on the first day of October, 1993. For the wastewater treatments, a full concentration of simulated regenerant wastewater was prepared, using a mixture of KCl, $MgCl_2$, and $CaCl_2$. For the irrigation treatments, the full concentration of simulated waste was diluted with regular tap water to 1/5, 1/10, and 1/20 dilutions. The 1/5 diluted wastewater had ionic concentrations of Ca^{2+}, 483; Mg^{2+}, 370; K^+, 165; and Cl^-, 2111 mg/liter. Concentrations of the 1/10 and 1/20 dilutions were approximately one-half and one-quarter of the 1/5 dilution, respectively. The control plots received only tap water. The turfgrass was irrigated with wastewater twice a week to the equivalent of 1/2 inch of water. During the dry and warm summer months, the turfgrass was irrigated once with tap water between wastewater irrigations.

Turfgrass clippings, irrigation leach water, and soil samples were collected and analyzed on a routine basis. Turfgrass clippings were weighed from each plot to determine difference in growth among the treatments. The grass clippings were also analyzed for the content of Ca, Mg, K, and Cl to see how much was removed from the system when the grass clippings were removed. The soil and leach water were also analyzed to determine the fate of the remaining nutrients.

The results of the experiment showed that there were no statistically significant differences between the four irrigation methods (tap water, 1/5, 1/10, and 1/20 dilution).

Chemical analysis of the plant tissue showed that there was no statistical difference in uptake of Ca^{2+}, Mg^{2+}, or K^+ between the four irrigation methods. Chloride uptake, however, was affected substantially by irrigation with the regenerant wastewater, and the differences were statistically significant. For example, in the summer months, the plants irrigated with tap water had a tissue chloride content of approximately 5 mg/g dry weight. In contrast, the plants irrigated with the 1/5 diluted wastewater had chloride levels of approximately 25 mg/g.

The fraction of applied chloride that was taken up by the turfgrass varied by species, by season, and by dilution level of the wastewater. In the spring, all five species of turfgrass were actively growing, and 60% of the applied chloride was taken up by the Kentucky bluegrass and the rye grass and removed with the clippings. At the same time, over 80% of the applied chloride was found in the clippings of tall fescue, Bermuda grass, and Zoysia grass. In contrast, during the winter months the cool-season species were only able to take up 10–30% of the applied chloride.

Analysis of the concentrations of soil exchangeable ions at the end of the experiment showed significant differences in the chloride and magnesium content within the various irrigation methods. Moreover, there was no statistical difference in the electrical conductivity between soils irrigated with tap water and with the 1/20 diluted wastewater. The higher-concentration 1/5 and 1/10 wastewater did, however, have a significant impact on soil conductivity.

The lack of any significant impact on turfgrass growth or on the uptake of calcium, magnesium, or potassium is encouraging. The uptake data for chloride as well as the impact of the 1/5 and 1/10 wastewater on soil conductivity suggest that appropriate care must be taken when irrigating turfgrass with diluted wastewater. The overall conclusion of the study by Wu et al. (1996) was therefore that if appropriately managed, turfgrass could be irrigated successfully with diluted regenerant wastewater, with minimal environmental impact.

Ornamental Plant Irrigation

Nine ornamental plant species commonly found in the California landscape and gardens were tested for their response to wastewater irrigation using water rich in Ca^{2+}, Mg^{2+}, K^+, and Cl^- (Wu et al., 1995). The nine species studied were azalea, Japanese boxwood, hydrangea, lace fern, nadina, pittosporum, hedge rose, rhaphiolepis, and jasmine. Three replicates were done for each species for irrigation with both wastewater and the control water.

Each of the plants was grown in a plastic pot 8 inches in diameter by 1 foot high. The soil was a mixture of equal amounts of sand, peat moss, and redwood sawdust. The plants were kept on a greenhouse bench under carefully controlled conditions of temperature and photon flux.

Simulated waste water was synthesized with concentrations of Ca^{2+} 400, Mg^{2+} 243, K^+ 117, and Cl^-, 1484 mg/liter; these concentrations are similar to what would be expected for KCl regenerant waste after dilution 1/10 with raw water. The plants were irrigated with the 1/10 dilution of the wastewater twice a week, plus once with a nutrient solution. The control plants were watered with tap water and nutrient solution. The treatments were sprinkled over the leaves of the plants, since it is likely that the same would happen with a home irrigation system. One inch (2.5 cm) of water was applied per irrigation.

After 12 weeks of irrigation treatment, each of the plants was inspected visually for symptoms of damage. Leaf samples were obtained for chemical analysis, and the plants were cut back to 2 inches above the soil surface. The irrigation experiments were then continued for another 12 weeks. At the end of the second 12-week interval, the plants were harvested, dried and weighed, and a second set of samples were obtained for chemical analysis. At the end of the irrigation experiment, soil samples were also obtained from each of the pots and analyzed for chemical composition.

Plant shoot regrowth (as the dry mass of plant tissue) was measured and reported for each plant species using both simulated wastewater as well as the control water. The ratio of regrowth mass to the control mass was termed the tolerance ratio. The results of Wu's experiments are reproduced in Table 9-1 (taken from Wu et al., *J. Environ. Hort.*, 13(2): 92–96, June 1995).

The tolerance data shown in Table 9-1 indicate that growth of lace fern was severely inhibited by the wastewater irrigation; this result is consistent with the visual observation of severe chlorosis. Two other species—pittosporum and nadina—showed significant reduction in

TABLE 9-1: Tolerance ratio for nine species of ornamental plants studied by Wu et al. (1995)

Plant Species	Tolerance Ratio (%)
Azalea	91
Japanese boxwood	103
Hydrangea	100
Lace fern	0.08
Nadina	25
Pittosporum	73
Hedge rose	58
Rhaphiolepis	119
Jasmine	115

regrowth but no visual indication of chlorosis. Most of the other species showed no significant impact from the wastewater treatment, with tolerance ratios in the range 91–119%.

Chemical analysis of the plant tissue showed that irrigation with the simulated wastewater led to a significant increase in chloride content relative to regular water treatment. The increase in chloride content varied with species and ranged from a 2-fold increase for rhaphiolepis to a 65-fold increase for the lace fern. There was a negative correlation between chloride uptake and tolerance ratio, indicating that greater chloride uptake was associated with reduced plant growth. The chloride impact, however, was tempered by a positive correlation between the tolerance ratio and the tissue calcium concentration, which suggests that calcium uptake can help to alleviate chloride toxicity.

The overall conclusion from Wu's study is that chloride appears to be the most significant factor when considering the use of diluted wastewater for irrigation of ornamental plants. Many species did not appear to suffer significant consequences. Severe impacts of the wastewater would only be expected for a small number of chloride-sensitive plants. Therefore, it is prudent to evaluate the sensitivity of landscape plants before irrigation with regenerant wastewater.

Studies such as the one reported by Wu are useful, since they should help the water treatment industry develop guidelines for the safe use/ disposal of KCl treated regenerant wastewater on turfgrass and ornamental plantings. Until definitive results come in, it is suggested that a minimum dilution of 75:1 regenerant wastewater with nontreated water. It is also recommended that grass clippings be removed at each mowing, and that the lawn be irrigated with nontreated water at least once between each application of diluted wastewater. Ornamental plants and fruits and vegetables are often more sensitive to dissolved salts than turfgrass, so a dilution of at least 200:1 is advisable if regenerant waste water is used in areas where these types of plants are growing.

IN CONCLUSION

Comparison of NaCl and KCl in the environment showed no single area of impact where NaCl was less harmful than KCl. The damage to soils from excess sodium is well documented by soil scientists, and high sodium levels are known to have a negative impact on crop yields and quality. In contrast, potassium is an essential crop nutrient that, appropriately applied, helps to improve crop performance.

In septic and sewage systems, sodium and potassium behave very similar. The major difference between them is that sodium loading to the environment (in sewage sludge or septic fields) risks damage to soils. Potassium appears to pose less of a risk for soil damage, especially when considering that potassium is efficiently removed by plants, thus reducing the risk of cumulative effects over time.

The bulk of regenerant salt used in water softening eventually reports to water bodies. The impacts of sodium and potassium on aquatic ecosystems are similar in many ways. Potassium has the benefit of being a plant nutrient, while sodium is not. Recent research has shown that in two important areas—inhibition of algae blooms and control of zebra mussels—potassium chloride has environmental performance superior to sodium chloride.

Sodium really has no redeeming value in our environment outside of salt water and brackish water ecosystems. If we develop alternatives to sodium chloride for water treatment, we should use them. Potassium chloride is a logical choice. It reduces sodium coming from the water softening system, reduces chloride levels, and serves as a fertilizer for plants. From an environmental perspective, it is an excellent alternative to sodium chloride.

REFERENCES

Ahmed, S., Swindale, L.D., and El-Swaify, S.A. Effects of adsorbed cations on physical properties of tropical red earths and tropical black earths. *J Soil Sci* 1969; 20:255–268.

Bajwa, M.S. and Josan, A.S. Prediction of sustained sodic irrigation effects on soil sodium saturation and crop yields. *Agric Water Management* 1989; 16:217–228.

Berthelet, M. and MacLeod, R.A. Effect of Na$^+$ concentration and nutritional Factors on the lag phase and exponential growth rates of the marine bacterium *Delays aesta* and other marine species. *Appl Environ Microbiol* 1989; 55:1754–1760.

Bower, C.A. Cation exchange equilibrium in soils affected by sodium Salts. *Soil Sci* 1959; 88:32–35.

Bower, C.A., Ogata, G., and Tucker, J.M. Sodium hazard of irrigation waters as influenced by leaching fraction and by precipitation or solution of calcium carbonate. *Soil Sci* 1968; 106:29–34.

Brooks, R.H., Bower, C.A., and Reeve, R.C. The effect of various exchangeable cations upon the physical conditions of soils. *Soil Sci Soc Am J* 1956; 20:325–327.

Brownell, P.F. and Nicholas, D.J. Some effects of sodium on nitrate assimilation and nitrogen fixation in *Anabaena Cylindrica. Plant Physiol* 1967; 42:915–921.

Carstensen, J., Henriksen, P., and Heiskanen, A. Summer algal blooms in shallow estuaries: Definition, mechanisms, and link to eutrophication. *Limnology Oceanography* 2007; 52(1):370–384.

Cecconi, S., Salazrand, A., and Martelli, M. The effect of different cations on the structural stability of some soils. *Agrochimica* 1963; 7:185–204.

Centre for Water Resources Studies *The Effect of Water Softeners on Onsite Wastewater Systems*. DalTech, Dalhousie University, On-Site Applied Research Program, 2001.

Chen, Y., Banin, A., and Borochovitch, A. Effect of potassium on soil permeability in relation to hydraulic conductivity. *Geoderma* 1983; 30.

Chiang, S.C., Radcliffe, D.E., Miller, W.P., and Newman, K.D. Hydraulic conductivity of three southeastern soils as effected by sodium, electrolyte concentration, and pH. *Soil Sci Soc Am J* 1987; 51:1293–1299.

Comeau, Y., Rabionwitz, B., Hall, K.J., and Olkham, W.K. Phosphate release and uptake in enhanced biological phosphorus removal from wastewater. *J Water Pollution Control Fed* 1987; 57:707–715.

Conte, F.S. *Pond Fertilization: Initiating an Algal Bloom*. Western Regional Aquaculture Center, Publication No. 104, 2000.

Corbitt, R.A. *Standard Handbook of Environmental Engineering*. McGraw-Hill, New York, 1990.

de March, B.G. Acute toxicity of binary mixtures of five cations to the freshwater amphipod *Gemmarus iacubtris. Can J Fisheries Aquatic Sci* 1988; 45:625–633.

Dewalle, F.B. and Schaff, R.M. Groundwater pollution by septic tank drainfields. *J Environ Eng* 1980; 106:631–646.

Dietz, T.H., Lynn, J.W., and Silverman, H. "Method for controlling bivalves such as zebra mussels", US Patent 5417987, 1995.

Fisher, S.W., Stromberg, P., Bruner, K.A., and Boulet, L.D. Molluscicidal activity of potassium to zebra mussel, *Dreissena polymorphia*: toxicity and mode of action. *Aquatic Toxicol* 1991; 20:219–234.

Gardner, W.R., Mayhugh, M.S., Goertzen, J.O., and Bower, C.A. Effect of electrolyte concentration and exchangeable sodium percentage on diffusivity of water in soils. *Soil Sci* 1959; 88:270–274.

Grasso, D., Strevett, K., and Pesari, H. Impact of sodium and potassium on environmental systems. *J Environ Syst* 1992; 22(4):297–323.

Henry, C.D., Moldenhauer, R.E., Engelbert, L.E., and Truog, E. Sewage effluent disposal through crop irrigation. *Sewage Indust Waste* 1954; 26:123–135.

Hesterberg, D. and Page, A.L. Critical coagulation concentrations of sodium and potassium illite as affected by pH. *Soil Sci Soc Am J* 1990; 54:735–739.

Imai, H. and Endoh, K. Potassium removal accompanied by enhanced biological phosphate removal. *J Fermentation Bioeng* 1990; 69:250–255.

Kappe, D.S. and Kappe, S. Algal growth exciters. *Water and Sewage Works* 1971; 118:245–248.

Keren, R. and Singer, M.J. Effect of low electolyte concentation on hydraulic conductivity of sodium/calcium-montmorillonite-sand system. *Soil Sci Soc Am J* 1988; 52:368–373.

Kincannon, D.F. and Gaudy, A.F. Some effects of high salt concentrations on activated sludge. *J Water Pollution Control Fed* 1966; 38:1148–1159.

Kirk, L.A. and Doelle, H.W. The effects of potassium and chloride ions on the ethanol fermentation of sucrose by Zymomonas mobilis 2716. *Appl Microbiol Biotechnol* 1992; 37:88–93.

Kresge, C.B. and Younts, S.E. Effect of various rates and freqencies of potassium application on yield and chemical composition of alfalfa and alfalfa-orchardgrass. *Agronomy J* 1962; 54:313–316.

Kugelman, I. and McCarty, P. Cation toxicity and stimulation in anaerobic waste treatment. *J Water Pollution Control Fed* 1965; 37:97–116.

Levy, G.J. and Torrento, R. *Soil Sci* 1995; 160:352–358.

Lewis, D.P., Piontkowski, J.M., Straney, R.W., Knowlton, J.J., and Neuhauser, E.F. Use of potassium for treatment and control of zebra mussel infestation in industrial fire protection water systems. *Fire Technol* 1997; 33(4):356–371.

Ludzack, F.J. and Noran, D.K. Tolerance of high salinities by conventional wastewater treatment processes. *J Water Pollution Control Fed* 1965; 37: 1404–1416.

Lui, D. The effect of sewage sludge land disposal on the microbiological quality of groundwater. *Water Res* 1982; 16:957–961.

MacLean, A.J. Soil retention and plant removal of potassium added at an excessive rate under field conditions. *Can J Soil Sci* 1977; 57:371–374.

Mancino, C.F. and Pepper, I.L. Irrigation of turfgrass with secondary sewage effluent: soil quality. *Agronomy J* 1992; 84:650–654.

Markevics, L.J. and Jacques, N.A. Enhanced secretion of glucosyltransferase by change in potassium ion concentration is accompanied by an altered pattern of membrane fatty acids in *Streptococcus salivarius*. *J Bacteriol* 1985; 161:989–994.

Martin, J.P. and Richards, S.J. Influence of exchangeable hydrogen and calcium and of calcium, potassium, and ammonium at different hydrogen levels on certain physical properties of soils. *Soil Sci Soc Am J* 1959; 23:335–336.

McCarty, P.L. Anaerobic waste treatment fundamentals. *Public Works* 1964; 9:92–96.

Mohleji, S.C. and Verhoff, F.H. Sodium and potassium loss effects on phosphorus transport in algal cells. *J Water Pollution Control Fed* 1980; 52:110–125.

Moore, M., (ed.) Pipeline. *National Small Flows Clearinghouse* 2001; 12(1): 1–7.

Munshower, F.F. *Practical Handbook of Disturbed Land Revegetation*, Lewis Publishers, Boca Raton, FL, 1994.

Nabors, M.W. *Introduction to Botany*, Pearson Education, San Francisco, CA, 2004.

Olafsson, J. Temperature structure and water chemistry of Caldera Lake, Oskjuvatn Island. *Limnology Oceanography* 1980; 25:779–788.

Onoda, T., Oshima, A., Fukunaga, N., and Nakatani, A. Effect of Ca and K on the intracellular pH of an *Escherichia coli* L-form. *J Gen Microbiol* 1992; 138:1265–1270.

Overman, A.R. and Leseman, W.G. Soil and groundwater changes under land treatment of wastewater. *Trans Am Soc Agric Eng* 1982; 25: 381–387.

Page, W. Examination of the role of Na in the physiology of the Na-dependent soil bacterium *Azotobacter salinestris*. *J Gen Microbiol* 1991; 137:2891–2899.

Palazzo, A.J. and Jenkins, T.F. Land application of wastewater: effect of soil and plant potassium. *J Environ Qual* 1979; 8:309–312.

Paredes, M.J., Moreno, E., Ramos-Cormenzana, A., and Martinez, J. Characteristics of soil after pollution with wastewater from olive oil extraction plants. *Chemosphere* 1987; 16:1557–1564.

Pupisky, H. and Shainberg, J. Salt effects on the hydraulic conductivity of a sandy soil. *Soil Sci Soc Am J* 1979; 43:429–433.

Ravina, I. The mechanical and physical behavior of Ca-clay and K-clay soil. *Ecol Stud* 1973; 4:131–140.

Reeve, R.C., Bower, C.A., Brooks, R.H., and Gschwend, F.B. A comparison of the effects of exchangeable Na and K upon physical conditions of soils. *Soil Sci Soc Am Proc* 1954; 18:130–132.

Rhoades, J.D. Quality of water for irrigation. *Soil Sci* 1972; 113:277–284.

Robertson, W.D., Cherry, J.A., and Sudicky, E.A. Groundwater contamination from two small septic systems on sand aquifers. *Groundwater GRWAAP* 1991; 29:82–92.

Seale, D.B., Boraas, M.E., and Warren, G.J. Effects of sodium and phosphate on growth of cyanobacteria. *Water Res* 1987; 21:625–631.

Shukla, B. and Rai, L.C. Potassium-induced inhibition of nitrogen and phosphorus metabolism as a strategy of controlling microcystis blooms. *World J Microbiol Biotechnol* 2007; 23(3):317–322.

Shukla, B. and Rai, L.C. Potassium-induced Inhibition of photosynthesis and associated electron transport chain of Microcystis: implication for controlling cyanobacterial blooms. *Harmful Algae* 2006; 5(2):184–191.

Siegrist, R.L. Soil clogging during subsurface wastewater infiltration as affected by effluent composition and loading rate. *J Environ Qual* 1988; 16:181–188.

Sommers, L.E. Chemical composition of sewage sludge and analysis of their potential use as fertilizers. *J Environ Qual* 1977; 6:225–232.

Sparks, D.L. *Environmental Soil Chemistry* Academic Press, 2003.

Sprott, D., Shaw, K., and Jarrell, K. Ammonia/potassium exchange in methanogenic bacteria. *J Biol Chem* 1984; 259:12602–12608.

Sprott, D., Shaw, K., and Jarrell, K. Methanogenesis and the K transport system are activated by divalent cations in ammonia-treated cells of *Methanospirillum hungatei*. *J Biol Chem* 1985; 260:9244–9250.

Strauss, R. The effects of different alkali salts on growth and mineral nutrients of *Lemma minor*. *Int Rev Gesamten Hydrobiol* 1976; 61:673–676.

Strobel, H. and Russell, J.B. Role of sodium in the growth of a ruminal selenomonad. *Appl Environ Microbiol* 1991; 57:1663–1668.

Tyler, E.J., Corey, R.B., and Olutu, M.U. *Potential Effects of Water Softener Use on Septic Tank Soil Absorption On-Site Waste Water Systems*, Water Quality Research Council, 1978.

Velasco-Molina, H.A., Swoboda, A.R., and Godfrey, C.L. Dispersion of soils of different mineralogy in relation to SAR and electrolyte concentration. *Soil Sci* 1971; 111:282–287.

Virginia Department of Game and Inland Fisheries *U.S. Fish and Wildlife Service, Final Environmental Assessment, Millbrook Quarry Zebra Mussel and Quagga Mussel Eradication*, Wildlife Diversity Division, Richmond, Virginia, 2005.

Volt, S.L. and Phinney, H.K. Mineral Requirements for the Growth of Anahaena Spiruides in vitro. *Can J Botany* 1968; 46:619–630.

Ward, A.K. and Wetzel, R.G. Sodium: some effects on blue-green algal growth. *J Physiol* 1975; 11:357–361.

Wildridge, P.J., Werner, R.G., Doherty, F.G., and Neuhauser, E.F. Acute toxicity of potassium to the adult zebra mussel *Dreissena polymorphia*. *Arch Environ Contamination Toxicol* 1998; 34(3):265–270.

Wu, L., Chen, J., Lin, H., Van Mantgem, P., Harivandi, M.A., and Harding, J.A. Effects of regenerant wastewater irrigation on growth and ion uptake of landscape plants. *J Environ Hort* 1995; 13(2):92–96.

Wu, L., Chen, J., Van Mantgem, P., and Ali Harivandi, M. Regenerant wastewater irrigation and ion uptake in five turfgrass species. *J Plant Nutrition* 1996; 19(12):1511–1530.

Yang, H. and Drake, H. Differential effects of sodium on hydrogen- and glucose-dependent growth of the acetogenic bacterium *Acetogenium kivui*. *Appl Environ Microbiol* 1990; 56:81–86.

CHAPTER 10

POTASSIUM AND HUMAN HEALTH

D. McCARRON

(Prepared for the Potash Corporation of Saskatchewan, Inc., June, 1991.)

OVERVIEW

This chapter is a summary of current knowledge on the role of dietary potassium in human health. Specifically, it focuses on the beneficial effects of adequate dietary potassium and the potentially hazardous effects of deficit potassium consumption. Beginning with a discussion of the physiologic aspects of the nutrient, the chapter includes brief descriptions of how potassium is regulated by the body and factors that may impair proper elimination of potassium. Medical conditions that are associated with excess potassium are discussed. The epidemiologic data linking deficit potassium intake with a variety of medical disorders are summarized, as are current findings from both laboratory and clinical studies regarding the established protective effect of potassium in the diet. Discussion of potential benefits and risks of increased dietary potassium concludes the chapter.

INTRODUCTION

Potassium is essential to the normal functioning of the human body, and it plays a variety of roles in accomplishing this. Stated most simply, potassium is a crucial element in sustaining the water content of cells, enabling muscles to contract, transmitting electrical impulses within the heart, and conducting nerve impulses. In the blood, potassium is involved in maintaining acid-alkali balance and in the conversion of carbohydrates into energy and amino acids, which are used by cells to develop and repair tissues.

Because the body is unable to produce potassium in quantities sufficient to meet its requirements, these must be satisfied by external sources, primarily the diet. Humans require an average daily intake of 2700 to 3500 mg potassium to maintain adequate total body potassium balance (National Research Council, 1989). The Canadian Recommended Nutrient Intake (RNI) for potassium is 70 mg per day per kilogram of body weight; the US Committee on Recommended Dietary Allowances (RDA) has set minimum potassium requirements at 2000 mg per day, with levels between 1875 and 5625 mg considered "safe and adequate daily intake" (National Research Council, 1989; Witney et al., 1990). The principal sources of potassium in the diet are 1) dairy products, which represent 35–40% of the total exposure to potassium; 2) potatoes; and 3) fruits and vegetables (McCarron et al., 1984; Table 10-1). These three sources represent approximately 80% of the total potassium ingested in a 24-hour period. Typically, fluid sources of potassium apart from milk, are few, with the exception of some fruit and vegetable juices.

CELLULAR PHYSIOLOGY OF POTASSIUM

Potassium is one of four major cations integral to normal electrolyte balance in all living organisms. In general terms, potassium provides the positive electrical charge inside all living cells of humans and lower vertebrate species. Potassium's positive intercellular charge is offset by the negative charges provided by phosphate organic anions and by sulfate and chloride. The sodium ion provides the counterbalancing charge across the cell membrane to the potassium ion. The volume of intercellular fluid is much larger than extracellular fluid, and therefore the quantity of potassium in the body is greater than that of sodium (Pierson et al., 1974). The unequal distribution between the quantity of sodium outside the cell as compared to that of the

TABLE 10-1: Leading Food Sources of Potassium (Witney et al., 1990)

FOOD	MEASURE	POTASSIUM (mg)	ENERGY (kcal)
Dairy Products			
Cottage Cheese, 2% fat	8 oz	217	205
Milk, whole	8 oz	370	150
Milk, skim	8 oz	406	86
Milk, dried, nonfat	8 oz	1160	244
Ice cream, vanilla	1 cup	412	223
Yogurt, plain, lowfat	1 cup	531	144
Fruits (fresh)			
Apple juice	1 cup	295	116
Apricots, raw	3 ea	313	51
Apricots, dried	5 ea	482	83
Avocado	6 oz	1097	305
Banana	1 med	451	105
Cantalope	½ med	825	94
Dates	10 ea	541	228
Orange juice, fresh	1 cup	496	111
Peaches, peeled	1 med	171	37
Peaches, dried	5 ea	648	311
Prunes, dried	5 ea	313	201
Prune juice, canned	1 cup	707	181
Raisins	1 cup	1089	435
Meat/Seafood			
Beef, ground, lean	3 oz	248	246
Beef, sirloin, broiled	3 oz	299	238
Chicken, breast, baked	3 oz	220	142
Clams, canned	3 oz	534	126
Salmon, broiled	3 oz	319	183
Scallops, breaded	3 oz	310	200
Trout, baked	3 oz	539	129
Vegetables/Legumes (fresh, cooked)			
Asparagus	8 oz	279	23
Beans, kidney	8 oz	658	208
Beans, green/snap	8 oz	373	44
Beans, navy	8 oz	669	259
Beans, pinto	8 oz	800	235
Broccoli	1 spear	490	42
Brussels sprouts	1 cup	491	60
Carrots	1 cup	354	70
Cauliflower	1 cup	400	30
Corn on the cob	1 med ear	192	83
Lentils	1 cup	731	231
Potato, baked	1 med	844	220
Spinach	1 cup	838	41
Tomato	1 cup	400	38
Tomato juice, canned	1 cup	537	42

potassium inside of the cell results in a net negative potential across all cell membranes (Shanes, 1958). Normal functioning of all nucleated cells is dependent upon normal maintenance of this membrane potential.

Intracellular potassium concentration is approximately 30 times greater than that of extracellular potassium (approximately 150 mEq/l inside the cell vs. 5 mEq/l outside). Several energy-dependent, membrane-associated transport processes have evolved in cells to sustain this concentration gradient (Catterall, 1988). These systems assure maintenance of potassium at the higher concentration inside the cell by transporting potassium actively against the gradient. Excess intracellular potassium elimination occurs readily due to a large outward direct gradient. However, these transport systems cannot compensate for low amounts of extracellular potassium, amounts insufficient to provide adequate intracellular potassium. Thus, potassium deficiency poses a much greater threat to the life of the cell, organ, and organism than does potassium excess (Sterns et al., 1981).

The principal, most serious manifestations of inadequate cellular potassium homeostasis are the failure of cardiovascular (Lown et al., 1951) and neurological functions (Knochel, 1982). Typically, potassium deficiency is manifested first by abnormal cell excitation because of the inability of the cell to maintain membrane potential (Weidmann, 1961). Life-threatening failure of cardiovascular and neurological regulation occurs with acute, severe potassium deficiency (Knochel, 1982). Less dramatic degrees of potassium deficiency manifest themselves as chronic disorders of essential physiologic systems, such as blood pressure regulation (Conn, 1965), acid-base excretion by the kidney (Adrogué and Madias, 1981; Lennon and Jemann, 1968), and insulin secretion and its peripheral actions (Dluh et al., 1972).

Thus there is no mineral or element more crucial to the normal intracellular regulation of physiologic function than potassium. Although several secondary protective mechanisms exist in the cell, organ, and body to assist in the prevention of intracellular potassium depletion, in the face of insistent or intense potassium insufficiency these mechanisms are not adequate to prevent the subsequent occurrence of chronic and/or acute medical conditions.

POTASSIUM BALANCE

Potassium is an *essential nutrient*, which, by definition, means that the body is unable to manufacture a potassium molecule; maintenance of potassium homeostasis is entirely dependent upon exogenous sources such as the diet. Almost all dietary potassium is absorbed in the proximal duodenum (Sterns et al., 1981), with only a minimal amount remaining in the final portions of the large bowel. The potassium is then transported into the blood and delivered to the cells of various organs as needed. Serum potassium levels rise once intracellular requirements are met, and the cell membrane-imbedded systems ensure extrusion of excess potassium.

The remaining potassium in the bloodstream, representing the smaller portion of total body potassium, is continually being removed from the blood as the latter moves across the glomeruli, or filtering mechanisms of the kidney. The filtered potassium is totally reabsorbed along the length of the early or proximal tubule segments of the kidney (Wright, 1981) and then returned to the blood. While potassium filtered in the proximal tubules is reabsorbed by the kidney, potassium in the late distal tubules is secreted back into the urine space.

Several important physiological factors determine appropriate renal reabsorption of potassium and then ultimate proper distal tubule secretion (Wright, 1981). Distal tubule secretion is largely dependent upon distal sodium delivery and the action of aldosterone, the mineral corticoid hormone secreted by the adrenal cortex. When dietary sodium and potassium intakes are adequate, aldosterone secretion by the adrenal cortex is turned off and normal quantities, reflecting the previous 24-hour intake of potassium, are secreted into the urine and eliminated from the body. In a healthy individual without intestinal, kidney, or specific endocrine disorders such as adrenal dysfunction, total urinary output of potassium equals the quantity of potassium ingested the previous day (Sterns et al., 1981). There are several pathological conditions in which potassium secretion may be inappropriately high and urinary potassium losses can exceed those quantities of the cation ingested in the previous 24-hour period. The most common of these conditions are extreme intravascular volume depletion and inadequate sodium intake.

In order to maintain extracellular volume and sodium balance, the kidney reabsorbs all available sodium and chloride, including that at distal nephron sites, and sacrifices potassium in that effort (Wright, 1981). Where volume depletion for prolonged periods of time occurs, such as with severe fluid and mineral losses or dietary sodium restriction coupled with diuretic use (Tannen, 1985; Seldin et al., 1984), urinary potassium losses may be significant and inappropriate in relation to total body potassium balance. Even where sodium intake is adequate, but chloride is not, urinary potassium losses may exceed dietary intake and depletion will result. This reflects the dependence of proximal tubule sodium reabsorption on the presence of chloride. When there is inadequate filtered chloride, increasing proportions of sodium that were not absorbed proximally are absorbed distally at the expense of increased secretion of potassium.

Systemic acidosis can also exaggerate urinary potassium losses (Adrogué et al., 1972). When acid production is increased, for whatever reason, cells buffer the greater hydrogen ion load by secreting potassium, which is then filtered by the kidney. Excessive amounts are lost to the urine because potassium delivery to the nephron may exceed the tubules' ability to reabsorb the cation (Hayslett and Binder, 1982). In the case of systemic alkalosis typically associated with volume depletion, urinary potassium losses may also exceed intake and subsequent potassium depletion will occur (Sterns et al., 1981). Dysfunctional absorption by the intestine of food and electrolytes can initiate excessive potassium loss; this is commonly seen with chronic diarrheal states (Krejs, 1987). However, short-term diarrhea, such as that related to viral infections seldom results in potassium deficiency.

DEFECTS IN POTASSIUM ELIMINATION

Potassium is eliminated from the body primarily by filtration of the blood by the kidney and the subsequent potassium secretion in the distal portions of the nephron into the urine, as described above. Because of the kidney's large capacity to secrete potassium, it is virtually impossible to become potassium overloaded if renal, or kidney, function is normal.

If, however, certain kidney functions are compromised, potassium elimination will likely be impaired. There are two aspects of renal potassium elimination that, if modified separately or together, will prevent normal potassium elimination from the body and increase the risk for potassium overload. Because the initial stage of potassium removal by the kidney is filtration of the blood, with the movement of the potassium ion across the glomerular membrane into the proximal tubule, any condition that significantly reduces the filtration of the blood will alter the body's ability to eliminate appropriate amounts of potassium on a daily basis (Sterns et al., 1981; Gonick et al., 1971; Schrier and Regal, 1972; Schon et al., 1974). Therefore, where significant renal failure has developed, there is a high probability of subsequent difficulties in maintaining appropriate normal levels of serum potassium and, ultimately, intracellular potassium. Whether genetic or acquired renal disease is the cause, when glomerular filtration rate is less than 15% of normal, patients are at risk of developing hyperkalemia (abnormally high serum potassium) and potassium overload; potassium intake should be carefully moni-tored in this situation.

It is important to note, however, that where renal function is reduced principally because of tubular dysfunction and not as a consequence of an impairment of glomerular filtration, there may actually be potassium wasting (Gonick, 1971). The inability of the tubules to reabsorb potassium offsets the reduction of glomerular filtration (Gonick et al., 1971; Schrier and Regal, 1972; Schon et al., 1974). In patients with tubular dysfunction (25–30% of all patients with significant renal disease), blood filtration continues unimpaired, but potassium reabsorption and secretion of filtered blood constituents are impaired. Filtered potassium passes through the nephron without being properly reabsorbed and then secreted in normal amounts. Thus significant renal disease is not always associated with the development of hyperkalemia or potassium overload, even if dietary intake of potassium is high.

The second condition under which renal dysfunction will result in potassium overload is where distal secretion of potassium in exchange for hydrogen or sodium is impaired. Since the distal secretion sites for potassium, which are critical for maintaining a normal potassium balance, are principally under hormone control, interference with these hormone-mediated transport systems, by either medical conditions or drug therapy, can result in elevated serum potassium and potassium overload.

There are acute conditions in which significant increases in serum potassium without increases in total body potassium may occur even when kidney function is normal. One such setting is where intracellular sources of potassium are rapidly introduced into the bloodstream. This most likely occurs in one of two situations. In patients with diabetes, there may be a shift of hydrogen ion into the cell in exchange for large quantities of potassium outside of the cell during the development of severe acidosis (Adrogué and Madias, 1981; Viberti, 1978). The second situation is that in which there has been sudden cell death in tissues (tissue necrosis) resulting in disturbances in the cell membrane. This subsequently causes the cell to discharge large quantities of intracellular potassium into the extracellular and, ultimately, the intravascular blood space (Lordon and Burton, 1972; Arseneau et al., 1973).

MEDICAL CONDITIONS RELATED TO POTASSIUM EXCESS

There is a potential for potassium overload in almost all forms of kidney disease. As described above, maintenance of a normal intracellular potassium level is critical to normal cell function and therefore to the function of all organ systems that comprise normal human physiology. If an individual is unable to eliminate the appropriate amounts of potassium daily, as may occur in the presence of kidney disease, then potassium builds up in the blood space. This impairs the ability of the cell to extrude intracellular potassium, and the sodium outside the cell is altered; that is, membrane potential is increased and cell function degenerates (Shanes, 1958; Adrian, 1956). This loss of normal cell responsiveness occurs because membrane potential is increased and factors that routinely trigger a cell action are no longer effective. Renal failure can be caused by a wide variety of both acquired and congenital conditions, all of which, apart from tubular disorders as described above, can induce hyperkalemia and potassium overload.

Pharmaceutical agents are another area in which there is potential risk of development of potassium overload and hyperkalemia. The actions of these medications can be either direct impairment of tubular ability to secrete potassium or indirect impairment by alteration of the other hormonal factors in the body that participate in controlling potassium secretion (Rimmer et al., 1987). Potassium excess does not generally develop from the use of these medications unless there is at least some degree of defect in kidney function. The specific drugs include nonsteroidal anti-inflammatory drugs (Tan et al., 1987; Zimran et al., 1985); potassium-sparing diuretics that directly block the secretion of potassium in the distal nephrons (Ponce et al., 1985); angiotensin-converting enzyme (ACE) inhibitors (Greenblatt, 1973; Textor et al., 1982; Burnakis and Mioduch, 1984); antiphypertensive drugs that impede the normal production of the hormone aldosterone; and possibly beta-blockers (Brown et al., 1983; Swenson, 1986; Lim et al., 1981). The latter group are administered for the treatment of high blood pressure, angina, and certain cardiac arrhythmias, and while these do not affect potassium elimination directly, they can alter the normal uptake of potassium in all body cells and, therefore, can produce hyperkalemia, particularly when potassium intake is increased.

Hyperkalemia and potassium overload are also risks for patients with diabetes mellitus. In this disease, acidosis develops when glucose control is not maintained normally (Nicolis et al., 1981), and hydrogen ions that are released during acidosis are taken up by the cell in exchange for potassium, which is extruded out of the cell. Another condition in which chronic hyperkalemia and potassium can occur is in the presence of adrenal hypofunction (hypoadrenalism) (Tuck and Mayes, 1980; Kigoshi et al., 1985). Similar to the effects of pharmaceutical agents that interfere with mineral-corticoid production, endogenous defects in aldosterone production, which are often present in hypoadrenalism or hypopituitarism, can result in hyperkalemia and potassium overload.

Potassium overload can also occur in certain acute medical situations. In hospital settings, when potassium is rapidly administered by intravenous methods, significant and potentially life-threatening hyperkalemia may result. The anesthetic succinylcholine, which is administered for surgical procedures, can cause potassium to leak out of the cells (Birch et al., 1969; Cooperman, 1979); this can become life-threatening when it occurs in conjunction with tissue destruction during an operative procedure. Rapid cell death, which most commonly occurs in tumor necrosis, that is, the rapid destruction of tumor tissue, and rhabdomyolysis, in which acute muscle damage results in skeletal muscle necrosis and the discharge of intracellular potassium into the blood space (Lordon, 1972; Arseneau et al., 1973; Knochel and Schlein, 1972).

Excess potassium does not lead to the development of any chronic medical disorders. The consequences of potassium overload are manifested in acute medical crises, the clinical expression of which is isolated solely to cardiac arrhythmias and sudden death. These occur because the normal membrane potential of the excitatory vascular smooth muscle cells that comprise the contractile forces of the heart and peripheral blood vessels are paralyzed. Contractility of the heart is regulated by the conduction of an electrical current through specific pathways of the heart, which then initiate cell membrane depolarization and an organized contraction of the right and left ventricles. If the cell membrane potential is not maintained normally, which it would not be with potassium overload, normal cardiac conduction to initiate contractility is disrupted. Consequently, there are acute spasmodic and disorganized contractions of the heart, or ventricular fibrillation, that prevent appropriate maintenance of blood pressure. This series of events will result in a cardiac event in which both cerebral and peripheral circulation cease.

MEDICAL CONDITIONS RELATED TO POTASSIUM DEFICIT

Epidemiologic Evidence

Two National Health and Nutrition Examination surveys (NHANES I and II) have been conducted throughout the population in the U.S. in the past two decades. Analysis of the databases acquired in these surveys has shown that as much as 25% of the American population does not consume the minimum daily recommended level of potassium. As previously stated, the U.S. RDA for potassium intake is a minimum of 2000 mg per day, and the primary sources of potassium in the diet are dairy products, potatoes, and other vegetables and fruits (McCarron et al., 1984; Table 10-1). Recent dietary studies report that almost 50% of all Americans eat no fruit and nearly 25% no vegetables on a given day.

Certain groups of individuals, for various reasons, include few foods in their diet that are considered potassium-rich or even adequate. In lower socioeconomic groups, financial restraints often limit consumption of dairy products; in some racial groups, these same products are avoided because of a high incidence of lactose intolerance. Responding to dietary trends, many individuals industrialized societies have reduced their consumption of certain foods in attempts to reduce caloric, sodium, and/or fat intake, and simultaneously and inadvertently reduced their intake of essential nutrients.

Epidemiologic data have indicated an association between deficit potassium and acute medical situations as well as chronic medical conditions. Potassium deficiency that develops over several weeks, that is, acute potassium deficiency, creates problems in settings of other chronic medical disorders such as cardiovascular disease, certain types of "salt-wasting" nephropathy, or acute intestinal dysfunction. The individual may be under extreme metabolic demands for adequate potassium intake as a result of the disease or of the medications used to treat it. If dietary intake of potassium is not maintained, it is possible for rapid potassium deficiency and cell dysfunction to develop. Typically, this situation clinically manifests as cardiac arrhythmias (Lown et al., 1951; Weidmann, 1961; Shapiro and Taubert, 1975), and in subjects predisposed because of established cardiovascular disease this can be life-threatening. Apart from cardiac arrhythmias, the acute onset of potassium deficiency manifests as neurologic dysfunction. This most typically presents as a neuromuscular spasm; the athlete exercising on a hot day without adequate repletion of the large cutaneous losses of potassium through perspiration typifies this situation.

One of the chronic medical conditions that has consistently been linked with long-term inadequate potassium intake is high blood pressure or hypertension. This relationship was first identified over 40 years ago, and since then a number of epidemiologic reports have demonstrated that individuals on diets that are routinely low in potassium are two to three times as likely to develop chronic increases in arterial pressure as individuals whose regular dietary potassium intake is within or above the recommended daily levels (McCarron et al., 1984; Reed et al., 1985; Harlan and Harlan, 1986; Liebman et al., 1986). It was first suggested that this relationship was related to the level of dietary sodium, but within the past five years multiple reports have identified a direct inverse correlation between the level of chronic dietary potassium ingestion and the probability of developing high blood pressure or the actual level of arterial pressure in a population (McCarron et al., 1984; Reed et al., 1985; Harlan and Harlan, 1986; Liebman et al., 1986). These studies have consistently noted a reduction in dietary potassium from dairy products as a principal reason why subjects are ingesting inadequate amounts of dietary potassium (McCarron et al., 1984; Reed et al., 1985). Based on data from the HANES (McCarron et al., 1984), the Honolulu Heart Project (Reed et al., 1985), and the Rancho Bernardo study (Ackley et al., 1983), one can project that potassium intake at a level of 44 mEq per day versus 75 mEq per day (1700 mg vs. 2900 mg per day) roughly doubles an individual's risk of developing hypertension.

The incidence of stroke in the United States has been similarly associated with inadequate intake of dietary potassium. After cancer and coronary disease, cerebrovascular accidents are the leading cause of premature morbidity and mortality in the United States and Canada. Analyses of the same population databases that were used to identify the link between low levels of potassium and elevated blood pressure have also demonstrated a significant reduction in stroke incidence and stroke-associated mortality with maintenance of adequate levels of dietary potassium intake. A major study in this area is that of Khaw and Barrett-Connor published in the *New England Journal of Medicine* (Khaw and Barrett-Connor, 1987). They characterized a two- to threefold increase in the risk of developing cerebrovascular disease and specifically stroke, fatal or nonfatal, in individuals consuming less than 50 mEq potassium per day versus those consuming more than 70 mEq.

There is also evidence from epidemiologic studies that, although not yet conclusive, indicates that individuals who are at risk of developing type II diabetes mellitus (insulin-resistant diabetes or late-onset adult diabetes) are more likely to consume diets of low potassium content (Durlach and Collery, 1984; Perez et al., 1977). The data relating low levels of dietary potassium to this endocrine disorder, which is especially common in an older population, are not presently of sufficient strength to make actual predictions of relative risks of developing type II diabetes or to assess the impact on the metabolic control of type II diabetes. A possible link between low dietary potassium consumption and obesity is also suggested indirectly by these findings (Perez et al., 1977; DeFronzo, 1988).

Investigators in a large, population-based Australian study assessed the dietary habits of patients with newly diagnosed colorectal cancer in comparison to those of healthy subjects and observed that persons who consumed diets high in potassium had a lower incidence of cancer of the large bowel (Kune et al., 1987). In a subsequent analysis of the data, in which statistical corrections were made for specific dietary factors, it was shown that potassium per se was not the sole factor affording the protective effect, but regular consumption of diets that included foods of high potassium content reduce the risk of developing colorectal cancer as compared to consumption of diets that do not contain high potassium foods (Kune et al., 1989).

In summary, many of the most common chronic diseases afflicting the adult population of industrialized societies are associated with low levels of dietary potassium intake. The evidence is strongest between low potassium intake and hypertension, stroke, and type II diabetes mellitus. In addition, the risk of life-threatening complications as a result of these disorders (for example, cardiac arrhythmias and diabetic acidosis) may be greatly increased by acute potassium depletion. Findings in epidemiologic studies indicate that chronic inadequate potassium intake is more common in older persons and in certain socioeconomic and ethnic groups. Thus those persons who are more prone to chronic medical disorders in Western cultures, such as the elderly, minorities, and the poor (McCarron et al., 1984; Reed et al., 1985; Harlan and Harlan, 1986; Liebman et al., 1986), are also those whose dietary potassium intake is consistently inadequate.

Laboratory Evidence

While epidemiologic studies can provide data regarding the incidence of disease within populations and relate this to reported dietary patterns, they are not able to establish cause-and-effect relationships. Diets that are low in potassium may also differ in the amounts of other nutrients they contain, each of which, independently and combined, may play a role in the pathogenesis of various disease states. Experimental (animal) and clinical (humans) studies are necessary to assess the specificity of potassium-induced changes in body function. In experimental or laboratory settings in which dietary factors can be carefully manipulated, increased potassium has been shown to have a protective effect on the development of chronic medical disorders associated with low dietary potassium intake in humans.

In studies performed to assess the effects of dietary potassium on blood pressure, several animal models of human hypertension have demonstrated reductions in the blood pressure increases typically described in these animals. Potassium supplementation has been shown to decrease blood pressure in the spontaneously hypertensive rat (SHR) (Workman, 1985), the Dahl salt-sensitive rat (Dahl et al., 1972), the DOCA-saline rat (Soltis et al., 1988), and the Goldblatt renovascular hypertension rat (Suzuki et al., 1981). This association holds for both acute dietary manipulations of sodium chloride in the case of the Dahl salt-sensitive rat and the DOCA saline model, and the more chronic gradual elevations in blood pressure observed in the genetic animal model, the SHR. In normotensive Spraque-Dawley rats who had developed hypertension on high-sodium diets, increased potassium in the diet lowered blood pressure and improved survival (Meneely and Ball, 1958; Meneely et al., 1961). Several theories have been proposed to explain this antihypertensive effect of dietary potassium, but none has been verified to date (Krishna, 1990).

Increased dietary potassium intake has been shown to afford protection against the onset of cerebrovascular episodes in Dahl salt-sensitive rats and stroke-prone SHR (Tobian et al., 1984, 1985; Tobian, 1986). The reduction in cerebral hemorrhage, stroke rate, and stroke-associated mortality in these animals subsequent to potassium increases in the diet has been disassociated from the degree of blood pressure decrease in the animal (Tobian et al., 1985). Animals in these studies developed renal arterial lesions (thickened walls and narrowed lumens) and focal tubular dilation. Vascular and tubular injury were reduced by both potassium nitrate and potassium chloride, suggesting a specific effect of potassium on the prevention of vascular wall damage that occurs with elevated intra-arterial pressure (Tobian et al., 1984, 1985).

Experimental studies of animal models of human type II diabetes mellitus and insulin resistance are as yet limited in number. One study of investigators well recognized for their work in the areas of both human and animal type II diabetes reported that reducing the study subject's level of dietary potassium intake exacerbates the condition, whereas increasing dietary potassium intake improves the peripheral actions of insulin (Mondon et al., 1968). As a result, manifestations of the inherent complications of diabetes are reduced in number and degree.

In summary, laboratory studies in animal models to examine the effects of dietary potassium on the regulation of arterial pressure, on the prevention of strokes and reduction of stroke-associated mortality, and on the control of carbohydrate metabolism in diabetes have all demonstrated a protective effect of increased dietary potassium intake consistent with a causal link between low levels of potassium intake and the development of these chronic disorders in animals and possibly in humans.

Clinical Evidence

In the disease states associated with low potassium intake, the associations provided by epidemiologic studies have been substantiated in animal models. The conclusive evidence, however, is best provided by studies of these direct relationships in humans. Similar to the favorable responses of increased potassium in the diet that have been reported in experimental models of hypertension, stroke, and diabetes, human clinical trials have shown a protective effect of potassium in these disease states.

In the case of human hypertension, comparable effects have been observed in controlled clinical trials. The response to potassium supplementation as a primary treatment of hypertension has been somewhat variable, with not all studies demonstrating a protective effect (Kaplan, 1988; Linas, 1991). However, the more rigorously controlled, double-blind, crossover trials have shown an antihypertensive effect of increasing potassium intake in subjects with mild to moderate essential hypertension (Siani, 1987; Grobbee et al., 1987; Ilmura et al., 1981; MacGregor et al., 1982; Svetkey et al., 1987; Lawton et al., 1990; Cappuccio and MacGregor, 1991). A meta-analysis published in the *Journal of Hypertension* reviewed data from 19 clinical studies of the effect of potassium on blood pressure and demonstrated, overall, that increased potassium significantly lowers both systolic and diastolic blood pressure (Cappuccio and MacGregor, 1991). This analysis also revealed that the longer the duration of increased potassium intake, the more pronounced the blood pressure-lowering effect.

Within specific groups or degrees of condition, the protective action of potassium has been even more evident. A leading example of this is the work of Kaplan and colleagues, who demonstrated that in diuretic-treated black females with hypokalemia (abnormally low serum potassium levels) and high blood pressure, modest increases in potassium intake resulted in significant reductions in blood pressure concurrent with moderate increases in serum potassium and urinary potassium excretion (Kaplan et al., 1985). This group is considered a high-risk population, namely, black females who are likely to chronically consume a diet low in of potassium content and whose risk is exaggerated by the known kaliuretic effects of thiazide drugs prescribed for the treatment of mild hypertension as discussed above.

Even in subjects with normal blood pressure (normotensive), increased dietary potassium has been shown to have a protective effect on blood pressure regulation (Khaw and Thom, 1982; Miller, 1987). In a study published in the *New England Journal of Medicine*, Krishna and coworkers reported the beneficial effect of potassium in normotensive subjects who were salt-sensitive (Krishna et al., 1989). For these individuals, blood pressure rises in response to high sodium intake regardless of baseline blood pressure levels. Two groups of normotensive men consumed their typical diets, maintaining their usual sodium chloride intake; potassium intake was manipulated at low or normal levels. On the low-potassium diets blood pressure increased significantly, whereas it did not change when normal amounts of potassium were included in the diet. This study indicates that adequate intake of potassium offsets the hypertensive effect of typical sodium intake in salt-sensitive persons, estimated to be roughly 30% of the general population, and 50% of persons with high blood pressure (Weinberger, 1986; Luft et al., 1989).

A study published in *Hypertension* identifies a connection between prenatal diets of mothers and the blood pressure of their infants for at least the first year of life (McGarvery et al., 1991). It has been demonstrated that blood pressures early in life are predictive of later-life blood pressures (Zinner et al., 1985; Schachter et al., 1984). McGarvery and colleagues reported that the intake of potassium by pregnant women was inversely and independently related to their infants' blood pressure at 6 and 12 months of age (McGarvery et al., 1991). The children of mothers who consumed the higher amounts of potassium during pregnancy had lower blood pressures during the first year of their lives than the children of mothers who consumed diets low in potassium during pregnancy. This protective effect of potassium on infant blood pressures was independent of other electrolytes in the diet.

Longitudinal clinical trials to assess the protective effect of dietary potassium against cerebrovascular disease in humans have not been conducted. It is presumed that the protective influence of dietary potassium against stroke and other cerebrovascular disorders is an effect that is afforded over a lifetime of chronic adequate potassium consumption. Therefore, conclusive documentation of this effect from clinical trials would require decades of observation and vast numbers of resident compliant subjects, as well as monumental financial support.

Adequate dietary potassium and thus the avoidance of potassium depletion in patients with type II diabetes mellitus has been shown to protect against the exacerbation of insulin resistance (Conn, 1965). Individuals who have both high blood pressure and type II diabetes and are being treated with thiazide diuretics for their hypertension are prone to potassium depletion. Degeneration in carbohydrate tolerance of glucose control will result in these patients if they are not maintained on higher daily levels of potassium (Helderman et al., 1983; Houston, 1988).

The risk of diuretic-induced cardiac arrhythmias, a potentially serious risk for patients receiving diuretic therapy for the treatment of hypertension or chronic cardiovascular conditions, has been shown to be lessened by increased potassium intake (Hollifield, 1984; Grimm, 1986; Hollifield, 1987). As discussed previously, diuretics impair potassium elimination, and if there is concurrent kidney dysfunction potassium overload can result from long-term diuretic use without appropriate potassium monitoring. The potassium and magnesium deficiencies that are associated with diuretic therapy have been hypothesized to result in an increased risk of ventricular arrhythmias and sudden death (Hollifield, 1984, 1987), and it has been demonstrated that these conditions can be normalized with resolution of the electrolyte deficiencies (Hollifield, 1984; Grimm, 1986).

The recent administration of thiazide diuretics (Tannen, 1985) and the simultaneous prescription of restricted sodium intake (Knochel, 1984), as given for certain cardiovascular conditions, can also predispose to potentially life-threatening problems. Extraordinary demands are placed on the kidney to secrete potassium in an effort to reabsorb the excess sodium chloride. Hypokalemia, cell dysfunction, and cardiac arrhythmias can develop rapidly in this setting, and these events are greatly amplified if the patient is also receiving cardiovascular drugs such as digitalis agents (Shapiro and Taubert, 1975). This combination of treatment is frequently prescribed in older persons with chronic congestive heart failure, and the potassium deficiency that may occur within a few days can result in sudden, life-threatening incidents.

Leg cramping, while not a life-threatening medical crisis, is a physically uncomfortable condition that has been shown to be relieved with increased dietary potassium. These muscle cramps are a common complication in elderly subjects, in patients who are receiving long-term thiazide therapy, and in persons who are restricting sodium intake as a means of lowering blood pressure.

There is a growing body of evidence regarding the critical role of potassium in the protection of skeletal mass (Lemann et al., 1989, 1991; Sebastian et al., 1990). Investigators comparing the effects of potassium bicarbonate on urinary calcium excretion and calcium balance to those of sodium bicarbonate reported that potassium reduced calcium excretion and improved body calcium balance while sodium had no effect (Lemann et al., 1989). A subsequent study by this group, published in *Kidney International*, compared the effects on calcium excretion of potassium bicarbonate and of potassium chloride with those of sodium bicarbonate and sodium chloride (Lemann et al., 1991); potassium administration in either form decreased urinary calcium excretion, whereas both forms of sodium were without effect. A study assessing the effects of potassium intake on serum phosphorus levels demonstrated that the level of dietary potassium determines the set point at which the kidney maintains serum phosphorus levels, and indicates that this phosphorus-regulatory mechanism then serves to modulate renal vitamin D production (Sebastian et al., 1990).

In summary, previous epidemiologic and experimental data linking dietary potassium deficits to the increased risk of developing a number of diseases and medical conditions have been validated in human clinical trials. While there is some variability in potassium responses in these trials, with not all individuals responding favorably, consistently significant findings of the beneficial, protective effects of dietary potassium have been reported by a number of independent groups of investigators. This evidence is most convincing in the areas of blood pressure regulation, control of carbohydrate metabolism in type II diabetes mellitus, and stroke incidence. Based on data from cross-sectional and intervention studies of hypertension and of strokes, the Canadian Consensus Report on Nonpharmacological Management of High Blood Pressure concludes that there is sufficient scientific evidence of the potential benefit of dietary potassium to recommend its increased intake for both hypertensive persons and healthy, normotensive individuals (Fodor and Chockalingam, 1990).

INDIRECT ROLE OF POTASSIUM IN HEALTH

As substantiated by the investigative work described above, there is a distinct relationship between several disease states and the intake of dietary potassium. The less direct effects of potassium must also be addressed. As is increasingly recognized by the scientific community (McCarron, 1991; Luft and McCarron, 1991), dietary nutrients are not consumed in isolation, and once ingested they do not function totally independent of other nutrients; they are interactive constituents of a total diet, and they express their physiological actions within integrated pathways in the body. In addition to the influence that dietary nutrients exert upon one another, they influence and are influenced by a myriad of other substances within the body. Further complicating these interactions, all of these components exist within multiple, interrelated regulatory systems, which themselves can be affected by genetic environmental, and lifestyle factors of the host body. Maintaining optimal human health is a complex, highly sophisticated process, dependent upon appropriate interactions between innumerable substances, systems, and conditions.

Therefore, while the lack or excess of a single nutrient may result in identifiable medical conditions, the necessity of an appropriate *balance* between all dietary nutrients cannot be overlooked. In addition to the role of potassium in the development of specific medical disorders, it must also be recognized that potassium is integral to the proper functioning of other nutrients in the body. Potassium and the other major cations (sodium, calcium, and magnesium) are required and utilized by the body in relation to one another, and inadequate or excessive amounts of any of these will alter the functional abilities of the others. While intricate regulatory systems exist to handle abnormal levels, these systems can be compromised in less than optimal health situations and/or if the abnormal levels are maintained over time. Thus, while inadequate dietary potassium intake has been *directly* implicated in the development of several specific diseases, its interactive role with other nutrients *indirectly* implicates its deficit intake in the development of any nutrient-related disorder.

SUMMARY

Potential Risks of Increased Dietary Potassium Intake

Potassium overload can occur in a very well-defined group of subjects with either renal disease or systemic or pharmacological conditions that impair potassium secretion by the kidney. The actual number of persons at risk of developing hyperkalemia and potassium overload probably represents less than 2% of the adult population. For the majority of these individuals, the primary source of potassium that places them at risk is supplements prescribed by physicians; diet sources of potassium are less likely to be the cause of hyperkalemia. In situations where hyperkalemia and cardiac consequences can evolve, excessive dietary potassium exposure may create a health risk to selected humans. The actual contribution of the potassium to that risk will be in proportion to the degree of the excess in intake.

The specific contribution to overall potassium intake from any given source will depend on the "density" of potassium from the source. Viewed in this context, water sources of potassium would, at best, represent less than 10% of excessive exposure to potassium. The actual health risk to humans of ingesting potassium from water, except under extraordinary circumstances, would appear to be close to zero. However, individuals should be informed of this exceedingly low risk and of the other sources and causes of potassium overload that are more likely to result in life-threatening medical conditions. Of all the potential sources for potassium excess, the greatest is the prescription of potassium supplements to patients who, for one or more of the reasons outlined above, would be unable to appropriately eliminate the excess potassium (Harrington et al., 1982; Lawson, 1974).

Adequate Dietary Potassium Intake

Because potassium is critical to the normal function of all cells, failure to maintain an adequate intake of dietary potassium to meet intracellular demands can result in the development of any of a number of chronic medical disorders. While low potassium intake is not the only factor in the etiology of these pathologic conditions, it has been shown to contribute to increased risk of their development. For most humans, the increased risk for developing these disorders is associated with relatively modest decreases in potassium, that is, 50 to 70 mEq per day as opposed to the recommended levels of 70 to 90 mEq per day (McCarron et al., 1984; Ackley et al., 1983). Thus even small increases in exposure to dietary potassium could result in significant benefits to the population. Because adequate amounts of dietary potassium are limited primarily to the food groups of dairy products and fruits and vegetables, it is common for individuals in our society to be ingesting chronically low levels of potassium. Any source of additional potassium intake, such as water, represents a potentially significant health benefit to the overwhelming majority of humans. While this additional source may only increase daily potassium intake by 10 to 15 mEq, this may be sufficient to offer important, long-term benefits to individuals and to society by ultimately reducing both the prevalence and the personal and economic costs of these common medical conditions (US Public Health Service, 1988; Table 10-2).

TABLE 10-2: The Burden of Diet-Related Illness in the U.S. Population

- **High Blood Pressure.** High blood pressure (hypertension) is a major risk factor for both heart disease and stroke. Almost 58 million people in the United States have hypertension, including 39 million who are under age 65. The occurrence of *hypertension* increases with age and is higher for black Americans (of which 38% are hypertensive) than for white Americans (29%).
- **Stroke.** Strokes occur in about 500,000 persons per year in the United States, resulting in nearly 150,000 deaths in 1987 and long-term disability for many individuals. Approximately 2 million living Americans suffer from stroke-related disabilities, at an estimated annual cost of more than $11 billion.
- **Coronary Heart Disease.** More than 1.25 million heart attacks occur each year (two-thirds of them in men), and more than 500,000 people die as a result. In 1985, illness and deaths from coronary heart disease cost Americans and estimated $49 billion in direct health care expenditures and lost productivity.
- **Cancer.** More than 475,000 persons died of cancer in the United States in 1987, making it the second leading cause of death in this country. During the same period, more than 900,000 new cases of cancer occurred. The costs of cancer for 1985 were estimated to be $22 billion for direct health care, $9 billion in lost productivity due to treatment or disability, and $41 billion in lost productivity due to premature mortality, for a total cost of $72 billion.
- **Diabetes mellitus.** Approximately 11 million Americans have diabetes, but almost half of them have not been diagnosed. In addition to the nearly 38,000 deaths in 1987 attributed directly to this condition, diabetes also contributes to an estimated 95,000 deaths per year from associated cardiovascular and kidney complications. In 1985, diabetes was estimated to cost $13.8 billion per year, or about 3.6% of total health care expenses.

Quoted from the 1988 *Surgeon General's Report on Nutrition and Health* (US Public Health Service, 1988).

REFERENCES

Ackley S, Barrett-Connor E, Suarez L. Dairy products, calcium, and blood pressure. *Am J Clin Nutr* 1983; 38:457.

Adrian RH. The effect of internal and external potassium concentration on the membrane potential of frog muscle. *J Physiol (Lond)* 1956; 133:631.

Adrogué HJ, Madias NE. Changes in plasma potassium concentration during acute acid-base disturbances. *Am J Med* 1981; 71:456.

Adrogué HJ, Lederer ED, Suki WN, Eknoyan G. Determinants of plasma potassium levels in diabetic ketoacidosis. *Medicine* 1972; 51:73.

Arseneau JC, Bagley CM, Anderson T, Canellos GP. Hyperkalaemia: a sequel to chemotherapy of Burkitt's lymphoma. *Lancet* 1973; 1:10.

Birch AA, Mitchell GD, Playford GA, Lang CA. Changes in serum potassium response to succinylcholine following trauma. *JAMA* 1969; 210:490.

Brown MJ, Brown DC, Murphy M. Hypokalemia from beta$_2$-receptor stimulation by circulating epinephrine. *N Engl J Med* 1983; 309:1414.

Burnakis TG, Mioduch JH. Combined therapy with captopril and potassium supplementation: a potential for hyperkalemia. *Arch Intern Med* 1984; 144:2371.

Cappuccio FP, MacGregor GA. Does potassium supplementation lower blood pressure: A meta-analysis of published trials. *J Hypertens* 1991; 9:465.

Catterall WA. Structure and function of voltage-sensitive ion channels. *Science* 1988; 242:50.

Conn JW. Hypertension, the potassium ion and impaired carbohydrate tolerance. *N Engl J Med* 1965; 273:1135.

Cooperman LH. Succinylcholine-induced hyyperkalemia in neuromuscular disease. *JAMA* 1979; 213:1867.

Dahl LK, Leitl G, Heine M. Influence of dietary potassium and sodium/potassium molar ratios on the development of salt hypertension. *J Exp Med* 1972; 136:318.

DeFronzo RA. Obesity is associated with impaired insulin-mediated potassium uptake. *Metabolism* 1988; 37:105.

Dluh RG, Axelrod L, Williams GH. Serum immunoreactive insulin and growth hormone response to potassium infusion in normal man. *J Appl Physiol* 1972; 33:22.

Durlach J, Collery P. Magnesium and potassium in diabetes and carbohydrate metabolism. *Magnesium* 1984; 3:315.

Fodor JG, Chockalingam A. The Canadian Consensus Report on Non-Pharmacological Management of High Blood Pressure. *Clin Exp Hypertens* 1990; 12:729.

Gonick HC, Kleeman CR, Rubini ME, Maxwell MH. Functional impairment in chronic renal disease. *Am J Med Sci* 1971; 261:281.

Greenblatt DJ, Koch-Weser J. Adverse reactions to spironolactone: a report from the Boston Collaborative Drug Surveillance Program. *JAMA* 1973; 224:40.

Grimm RH Jr. The drug treatment of mild hypertension in the Multiple Risk Factor Intervention Trial: a review. *Drugs* 1986; 31(Suppl 1):13.

Grobbee DE, Hofman A, Roelandt JT, Boomsma F, Schalekamp MA, Valkenburg HA. Sodium restriction and potassium supplementation in young people with mildly elevated blood pressure. *J Hypertens* 1987; 5:115.

Harlan WR, Harlan LC. An epidemiological perspective on dietary electrolytes and hypertension. *J Hypertens* 1986; 4(Suppl 5):S334.

Harrington JT, Isner JM, Kassirer JP. Our national obsession with potassium. *Am J Med* 1982; 73:155.

Hayslett JP, Binder JH. Mechanism of potassium adaptation. *Am J Physiol* 1982; 243:F103.

Helderman JH, Elahi D, Andersen DK, et al. Prevention of the glucose intolerance of thiazide diuretics by maintenance of body potassium. *Diabetes* 1983; 32:106.

Hollifield JW. Magnesium depletion, diuretics, and arrhythmias. *Am J Med* 1987; 82(Suppl 3A):30.

Hollifield JW. Potassium and magnesium abnormalities: diuretics and arrhythmias in hypertension. *Am J Med* 1984; Nov:28.

Houston MC. The effects of antihypertensive drugs on glucose intolerance in hypertensive nondiabetics and diabetics. *Am Heart J* 1988; 115:640.

Ilmura O, Kijuma T, Kikuchi K, et al. Studies on the hypotensive effect of high potassium intake in patients with essential hypertension. *Clin Sci* 1981; 61:77s.

Kaplan NM, Carnegie A, Raskin P, et al. Potassium supplementation in hypertensive patients with diuretic-induced hypokalemia. *N Engl J Med* 1985; 312:746.

Kaplan NM. Calcium and potassium in the treatment of essential hypertension. *Semin Nephrol* 1988; 8:176.

Khaw K-T, Barrett-Connor E. Dietary potassium and stroke-associated mortality. *N Engl J Med* 1987; 316:235.

Khaw K-T, Thom S. Randomised double-blind cross-over trial of potassium on blood pressure in normal subjects. *Lancet* 1982; ii:1127.

Kigoshi T, Morimoto S, Uchida K. Unresponsiveness of plasma mineralocorticoids to angiotensin II in diabetic patients with asymptomatic normoreninemic hypoaldosteronism. *J Lab Clin Med* 1985; 105:195.

Knochel JP, Schlein EM. On the mechanism of rhabdomyolysis in potassium depletion. *J Clin Invest* 1972; 51:1750.

Knochel JP. Diuretic-induced hypokalemia. *Am J Med* 1984; 77(Suppl 5A):18.

Knochel JP. Neuromuscular manifestations of electrolyte disorders. *Am J Med* 1982;72:521.

Krejs GJ. VIPoma syndrome. *Am J Med* 1987; 82(Suppl 5B):37.

Krishna GG, Miller E, Kapoor S. Increased blood pressure during potassium depletion in normotensive men. *N Engl J Med* 1989; 320:1177.

Krishna GG. Effect of potassium intake on blood pressure. *J Am Soc Nephrol* 1990; 1:43.

Kune GA, Kune S, Watso LF. Dietary sodium and potassium intake and colorectal cancer risk. *Nutr Cancer* 1989; 12:351.

Kune S, Kune GA, Watson LF. Case-control study of dietary etiological factors: the Melbourne Colorectal Cancer Study. *Nutr Cancer* 1987; 9:21.

Lawson DH. Adverse reactions to potassium chloride. *Q J Med* 1974; 43:433.

Lawton WJ, Fitz AE, Anderson EA, Sinkey CA, Coleman RA. Effect of dietary potassium on blood pressure, renal function, muscle sympathetic nerve activity, and forearm vascular resistance and flow in normotensive and borderline hypertensive humans. *Circulation* 1990; 81:173.

Lemann J Jr, Gray RW, Pleuss JA. Potassium bicarbonate, but not sodium bicarbonate, reduces urinary calcium excretion and improves calcium balance in healthy men. *Kidney Int* 1989; 35:688.

Lemann J Jr, Pleuss JA, Gray RW, Hoffman RG. Potassium administration [increases] reduces and potassium deprivation [reduces] increases urinary calcium excretion in healthy adults. *Kidney Int* 1991; 39:973. [Note: words in title were reversed and misprinted; bracketed words are correct.]

Lennon EJ, Jemann J Jr. The effect of a potassium-deficient diet on the pattern of recovery from experimental metabolic acidosis. *Clin Sci* 1968; 34:365.

Liebman M, Chopin LF, Carter E, et al. Factors related to blood pressure in a biracial adolescent female population. *Hypertension* 1986; 8:843.

Lim M, Linton RAF, Wolff CB, Band DM. Propranolol, exercise and arterial plasma potassium. *Lancet* 1981; 2:591.

Linas SL. The role of potassium in the pathogenesis and treatment of hypertension. *Kidney Int* 1991; 39:771.

Lordon RE, Burton JR. Post-traumatic renal failure in military personnel in Southeast Asia. *Am J Med* 1972; 53:137.

Lown B, Salzberg H, Enselberg CD, Weston RE. Interrelationship between potassium metabolism and digitalis toxicity in heart failure. *Proc Soc Exp Biol Med* 1951; 76:797.

Luft FC, McCarron DA. Heterogeneity of hypertension: the diverse role of electrolyte intake. *Ann Rev Med* 1991; 42:347.

Luft FC, Miller JZ, Lyle RM, Melby CL, McCarron DA, Weinberger MH, Morris CD. The effect of dietary interventions to reduce blood pressure in normal humans. *J Am Coll Nutr* 1989; 8:495.

MacGregor GA, Smith SJ, Markandu ND, Banks RA, Sagnella GA. Moderate potassium supplementation in essential hypertension. *Lancet* 1982; ii:567.

McCarron DA, Morris CD, Henry HJ, Stanton JL. Blood pressure and nutrient intake in the United States. *Science* 1984; 224:1392.

McCarron DA. A consensus approach to electrolytes and blood pressure: could we all be right? *Hypertension* 1991; 17(Suppl I):I170.

McGarvey ST, Zinner SH, Willett WC, Rosner B. Maternal prenatal dietary potassium, calcium, and infant blood pressure. *Hypertension* 1991; 17:218.

Meneely GR, Ball COT. Experimental epidemiology of chronic sodium chloride toxicity and the protective effect of potassium chloride. *Am J Med* 1958; 25:713.

Meneely GR, Lemley-Stone J, Darby WJ. Changes in blood pressure and body sodium of rats fed sodium and potassium chloride. *Am J Cardiol* 1961; 8:527.

Miller JZ, Weinberger MH, Christian JC. Blood pressure response to potassium supplementation in normotensive adults and children. *Hypertension* 1987; 10:437.

Mondon CE, Burton SD, Grodsky GM, Ishida T. Glucose tolerance and insulin response of potassium-deficient rat and isolated liver. *Am J Physiol* 1968; 215:779.

National Research Council. *Recommended Dietary Allowances*. Washington DC: National Academy Press, 1989.

Nicolis GL, Kahn T, Sanchez A, Gabrilove JL. Glucose-induced hyperkalemia in diabetic subjects. *Arch Intern Med* 1981; 141:49.

Perez GO, Lespier L, Knowles R, Oster JR, Vaamonde CA. Potassium homeostasis in chronic diabetes mellitus. *Arch Intern Med* 1977; 137:1018.

Pierson RN Jr, Lin DHY, Phillips RA. Total-body potassium in health: effects of age, sex, height, and fat. *Am J Physiol* 1974; 225:206.

Ponce SP, Hennings AC, Madias NE, Harrington JT. Drug-induced hyperkalemia. *Medicine* 1985; 64:357.

Reed D, McGee D, Yano K, Hankin J. Diet, blood pressure, and multicollinearity. *Hypertension* 1985; 7:405.

Rimmer JM, Horn JF, Gennari FJ. Hyperkalemia as a complication of drug therapy. *Arch Intern Med* 1987; 147:867.

Schachter J, Kuller LF, Peffetti C. Blood pressure during the first five years of life: relation to ethnic group (black or white) and to parental hypertension. *Am J Epidemiol* 1984; 119:541.

Schon DA, Silva P, Hayslett JP. Mechanism of potassium exretion in renal insufficiency. *Am J Physiol* 1974; 227:1323.

Schrier RW, Regal EM. Physiological role of aldosterone in sodium, water and potassium metabolism in chronic renal disease. *Kidney Int* 1972; 1:156.

Sebastian A, Hernandez RE, Portale AA, Colman J, Tatsuno J, Morris RC Jr. Dietary potassium influences kidney maintenance of serum phosphorus concentration. *Kidney Int* 1990; 37:1341.

Seldin D, Welt L, Cort J. The role of sodium salts and adrenal steroids in the production of hypokalemic alkalosis. *Yale J Biol Med* 1984; 246:F609.

Shanes AM. Electrochemical aspects of physiological and pharmacological action in excitable cells. II. The action potential and excitation. *Pharmacol Rev* 1958; 10:165.

Shapiro W, Taubert K. Hypokalaemia and digoxin-induced arrhythmias. *Lancet* 1975; 2:604.

Siani A, Strazzullo P, Russo L, et al. Controlled trial of long term oral potassium supplements in patients with mild hypertension. *Br Med J* 1987; 294:1453.

Soltis EE, Iloeje E, Katovich MJ. Dietary potassium blood pressure and peripheral adrenergic responsiveness in the deoxycorticosterone acetate-salt rat. *Clin Exp Hypertens* 1988; A10:447.

Sterns RH, Cox M, Feig PU, Singer I. Internal potassium balance and the control of the plasma potassium concentration. *Medicine* 1981; 60:339.

Suzuki H, Kondon K, Saruta T. Effect of potassium chloride on the blood pressure in two-kidney, one-clip Goldblatt hypertensive rats. *Hypertension* 1981; 3:566.

Svetkey LP, Yarger WE, Feussner JR, DeLong E, Klotman PE. Double-blind placebo-controlled trial of potassium chloride in the treatment of mild hypertension. *Hypertension* 1987; 9:444

Swenson ER. Severe hyperkalemia as a complication of timolol: a topically applied beta-adrenergic antagonist. *Arch Intern Med* 1986; 146:1220.

Tan SY, Shapiro R, Franco R, et al. Indomethacin-induced prostaglandin inhibition with hyperkalemia: a reversible cause of hyporeninemic hypoaldosteronism. *Ann Intern Med* 1987; 90:783.

Tannen RL. Diuretic-induced hypokalemia. *Kidney Int* 1985; 28:988.

Textor SC, Bravo EL, Fouad FM, Tarazi RC. Hyperkalemia in azotemic patients during angiotensin-converting enzyme inhibition and aldosterone reduction with captopril. *Am J Med* 1982; 73:719.

Tobian L, Lange J, Ulm K, Wold L, Iwai J. Potassium reduces cerebral hemorrhage and death rate in hypertensive rats, even when blood pressure is not lowered. *Hypertension* 1985; 6:I110.

Tobian L, MacNeil D, Johnson MA, Ganguli MC, Iwai J. Potassium protection against lesions of the renal tubules, arteries, and glomeruli and nephron loss in salt-loaded hypertensive Dahl rats. *Hypertension* 1984; 6:1170.

Tobian L. High-potassium diets markedly protect against stroke deaths and kidney disease in hypertensive rats, an echo from prehistoric days. *J Hypertens* 1986; 4(Suppl 4):S67.

Tuck ML, Mayes DM. Mineralocorticoid biosynthesis in patients with hyporeninemic hypoaldosteronism. *J Lab Clin Endocrinol Metab* 1980; 50:341.

US Public Health Service. *The Surgeon General's Report on Nutrition and Health.* DHHS (PHS) Publication No. 88-50211. Washington DC: US GPO, 1988.

Viberti GC. Glucose-induced hyperkalaemia: a hazard for diabetics? *Lancet* 1978; 1:690.

Weidmann S. Membrane excitation in cardiac muscle. *Circulation* 1961; 24:499.

Weinberger MH, Miller JZ, Luft FC, Grim CE, Fineberg NS. Definitions and characteristics of sodium sensitivity and blood pressure resistance. *Hypertension* 1986; 8(Suppl II):II127.

Whitney EN, Hamilton EMN, Rolfes SR. *Understanding Nutrition.* St. Paul MN: West Publishing, 1990.

Workman ML, Paller MS. Cardiovascular and endocrine effects of potassium in spontaneously hypertensive rats. *Am J Physiol* 1985; 249 (*Heart Circ Physiol* 18):H907.

Wright FS. Potassium transport by successive segments of the mammalian nephron. *Fed Proc* 1981; 40:2398.

Zimran A, Kramer M, Plaskin M, Hershko C. Incidence of hyperkalaemia induced by indomethacin in a hospitalized population. *Br Med J* 1985; 291:107.

Zinner SH, Rosner B, Oh W, Kass EH. Significance of blood pressure in infancy: familial aggregation and predictive effect on later blood pressures. *Hypertension* 1985; 7:411.

CHAPTER 11

ONGOING RESEARCH

Research into the effective use of KCl as a softener regenerant contin-ues in many laboratories around the world. A selection of research results from recent experiments is included in the present chapter.

IRON AND MANGANESE REMOVAL

Iron and manganese are troublesome elements found in some ground-waters (Henke, 1995). The quantities of these constituents in house-hold water must be quite low—less than 0.30 ppm for iron and 0.05 ppm for manganese. Higher levels of iron and manganese will result in stain-ing of laundry and porcelain fixtures. Iron will also impart an unpleas-ant taste to drinking water, which can be detected by some people at levels as low as 0.10 ppm.

Iron can be very difficult to treat because it is found in many differ-ent forms. In some cases it can be colloidal (i.e., very small solid par-ticles) iron oxide, or other iron-containing minerals. Iron can also be chemically attached to organic acids commonly found in groundwaters. In other cases, the iron is dissolved in the water—the water will be

Water Softening with Potassium Chloride: Process, Health, and Environmental Benefits, by William Wist, Jay H. Lehr, and Rod McEachern
Copyright © 2009 by John Wiley & Sons, Inc.

"clear when drawn." In this case the iron can be successfully removed by ion exchange, along with softening, provided that some care is taken in the management of the water softener:

— The water must not contact air before the softener. If water containing soluble iron contacts air, the iron will be oxidized to the ferric form and promptly precipitate as reddish iron oxides. These precipitates can plug and foul water softeners, as well as staining laundry.
— The softener needs to be regenerated more frequently than with iron-free water, i.e., before the system reaches softening capacity exhaustion.
— The level of iron plus manganese is acceptably low—most manufacturers recommend that the level of these two elements combined is less than 5 ppm.

Since iron and manganese can be problematic to water softening systems, research has recently been done to compare the performance of KCl and NaCl regenerants when treating influent water rich in both iron and manganese.

Two influent water samples were processed in the experiment. The first influent was standard hard water (7.56 gpg hardness as $CaCO_3$) and free of dissolved iron or manganese. The second influent water contained iron and manganese and was synthesized with the composition:

Calcium and magnesium hardness = 7.56 grains/gallon
Soluble iron = 11.3 ppm
Soluble manganese = 2.25 ppm
pH = 2.0

The influent water was made acidic, by addition of the appropriate amount of HCl, to prevent oxidation and precipitation of iron hydroxides.

The capacity of the ion exchange system was measured with the standard (Fe and Mn free) influent after regeneration—first with NaCl and then with KCl. The end point for these experiments was deemed to be an effluent with a hardness of 1.0 grains/gallon.

In a second experiment, the capacity of the ion exchange system was measured with Fe- and Mn-containing influent after regeneration—first with NaCl and then with KCl. The end point for the iron/manganese experiments was deemed to be 0.3 ppm Fe or 0.05 ppm Mn

TABLE 11-1: Capacity of a cation exchange resin (grains/ft³ with a regenerant dosage of 15 pounds) comparing regeneration with NaCl versus KCl and influent water containing iron and manganese

Regenerant	Influent Composition			Capacity (grains/ft³)
	Hardness (gpg)	Fe (ppm)	Mn (ppm)	
NaCl	7.56	0.0	0.0	29,855
KCl	7.56	0.0	0.0	29,332
NaCl	7.56	11.3	2.25	27,168
KCl	7.56	11.3	2.25	27,446

(whichever came first) in the effluent. The results from the capacity experiments are summarized in Table 11-1.

The capacity experiments shown in Table 11-1 confirm the conclusions presented in Chapter 8—that there is no significant difference in the capacity when regenerated with either KCl or NaCl at high regenerant levels. In addition, Table 11-1 shows the impact of iron and manganese on softener capacity. The average capacity with no Fe or Mn in the influent was $(29,855 + 29,332)/2 = 29,594$ grains/ft³. The presence of Fe and Mn in the influent reduced the average capacity to $(27,168 + 27,446)/2 = 27,307$ grains/ft³. Expressed as a percentage, the reduction in softener capacity was:

$$\frac{(29,594 - 27,307)}{29,594} \times 100\% = 7.7\%$$

The long-term effectiveness of iron and manganese removal was also studied by putting the ion exchange resin through a series of eight service/regeneration cycles with NaCl regenerant and a second series of eight cycles with KCl regenerant. For each of the cycles, the total volume of influent was fixed at 120 gallons, since it was found in the initial experiments that this volume of water could be treated before the effluent reached 0.3 ppm Fe or 0.05 ppm Mn. Degradation of resin performance was monitored by measuring the composition of the final effluent—after 120 gallons of water had been softened. The results showed that the Fe and Mn content of the final effluent started to rise significantly after repeated service cycles with Fe- and Mn-laden influent, as shown in Table 11-2 and illustrated in Figures 11-1 and 11-2. It should be noted, however, that there was no substantial difference in performance between KCl and NaCl regenerants.

TABLE 11-2: Iron and manganese content of the final effluent (after treating 120 gallons of water) for repeated softening cycles, using either NaCl or KCl as regenerant

Regenerant	Service Cycle	Final Effluent Composition	
		Fe (ppm)	Mn (ppm)
NaCl	1	0.11	0.06
NaCl	2	0.11	0.03
NaCl	3	0.11	0.04
NaCl	4	0.11	0.04
NaCl	5	0.11	0.07
NaCl	6	0.11	0.06
NaCl	7	0.16	0.07
NaCl	8	0.22	0.13
KCl	1	0.11	0.06
KCl	2	0.16	0.06
KCl	3	0.14	0.06
KCl	4	0.15	0.06
KCl	5	0.17	0.09
KCl	6	0.20	0.09
KCl	7	0.22	0.10
KCl	8	0.20	0.12

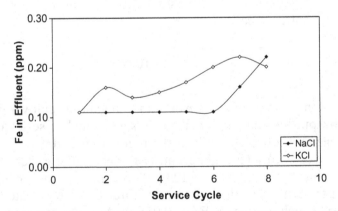

FIGURE 11-1 Iron content of the final (i.e., after treating 120 gallons) effluent after repeated service cycles using either NaCl or KCl as a regenerant. In all cases the influent iron was 11.3 ppm.

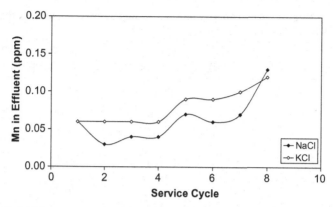

FIGURE 11-2 Manganese content of the final (i.e., after treating 120 gallons) effluent after repeated service cycles using either NaCl or KCl as a regenerant. In all cases the influent manganese was 2.25 ppm.

GRADE OF KCl REGENERANT

One of the barriers to increased use of KCl regenerant in water soften-ing is the cost. Commercial production of KCl is more complex, and therefore more expensive, than for NaCl. The cost of commercial KCl varies, however, according to the product grade—the composition can vary from 95% to 99.9% KCl, and the price increases with purity. Some recent research has therefore been done to determine whether lower-grade KCl could be used for water softening.

Commercial KCl sold for water softening is a recrystallized (white) product consisting of KCl and minor amounts of NaCl. The KCl product sold for water softening is typically 99.8% pure KCl, with guaranteed less than 0.2% NaCl; this product is known commercially as WSM 0.2. Another recrystallized product, known as WSM 1.0, has a lower product grade—it is typically 99.0% pure KCl, and guaranteed less than 1.0% NaCl. Aside from NaCl content, the level of other impurities in these two products is very low and does not vary much between the grades.

Research was done to determine whether the NaCl content of com-mercial KCl regenerant had a significant impact on performance. Experiments were done with commercial NaCl, WSM 0.2, and WSM 1.0 to measure the regeneration capacity as well as the percentage of unused regenerant. Experiments were performed as per the standard protocol described in Appendix 1 with regenerant dosages of 4, 8, and 15 pounds per cubic foot of resin. Calculations of capacity as well as the percentage of unused regenerant were done with the methodology described in Chapter 8. The results of the experiments are shown in

TABLE 11-3: Regeneration capacity and percentage unused regenerant for NaCl, WSM 0.2 (99.8% KCl), and WSM 1.0 (99.0% KCl)

Regenerant Type	Regenerant Dosage (lbs)	US Gallons Softened	Capacity (grains/ft^3/lb)	Percentage Unused
NaCl	4	653	4920	18.0
NaCl	8	980	3691	38.4
NaCl	15	1197	2392	60.1
WSM 0.2	4	627	4686	0.3
WSM 0.2	8	975	3648	22.4
WSM 0.2	15	1161	2323	50.6
WSM 1.0	4	604	4521	3.8
WSM 1.0	8	973	3651	22.3
WSM 1.0	15	1147	2310	50.9

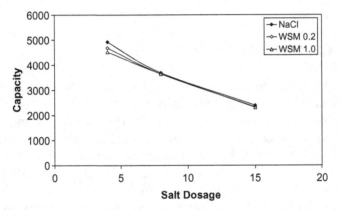

FIGURE 11-3   Regeneration capacity (grains per cubic foot of resin per pound of regenerant) for NaCl, WSM 0.2 (99.8% KCl), and WSM 1.0 (99.0% KCl) as a function of regenerant dosage (pounds).

Table 11-3 and illustrated in Figures 11-3 and 11-4. Data shown in Table 11-3 are the average of four replicate experiments.

The data provided in Table 11-3 and illustrated in Figure 11-3 show clearly that there are no substantial differences in performance between WSM 0.2 and WSM 1.0. There are possibly some minor differences between the capacity of WSM 0.2 and WSM 1.0 at the lowest regenerant level (4 pounds) but nothing significant enough to impact softener operation.

Figure 11-4 reinforces the conclusions discussed in Chapter 8—that KCl is nearly equivalent to NaCl on a mass basis, despite its higher molecular weight, because less of the KCl is unused during regeneration. The differences in performance between WSM 0.2 and WSM 1.0

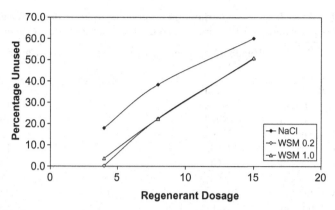

FIGURE 11-4   Percentage of unused regenerant for NaCl, WSM 0.2 (99.8% KCl), and WSM 1.0 (99.0% KCl) as a function of regenerant dosage (pounds).

were very minor, and only noticeable at the very lowest regenerant dosage.

The overall conclusion from the research into product grade was that there is no significant difference between 99.0% and 99.8% pure KCl (WSM 0.2 vs. WSM 1.0). The lower-purity WSM 1.0 could be substituted for WSM 0.2 at the same regeneration dosages with no impact on softening performance.

COCURRENT VERSUS COUNTERCURRENT REGENERATION

All the research described in Chapters 7 and 8 comparing the performance of KCl and NaCl regenerants was performed by using water softeners with countercurrent regeneration (Chapter 4). Cocurrent regeneration, however, is also commonly found in household softeners, especially among older units. A brief test was therefore performed to ensure that the conclusions reached in Chapters 7 and 8 were still valid with a cocurrent regeneration cycle.

Cocurrent and countercurrent softeners with approximately the same amount of resin were purchased from the same manufacturer. A second cocurrent unit was also purchased from a different manufacturer. The protocol developed for comparison of KCl and NaCl regenerants (Appendix 1) was used, with regenerant dosages of 4, 6, 8, and 15 pounds per cubic foot of resin. In the cocurrent versus countercurrent testing, only duplicate determinations of capacity were made for each regenerant dosage. The results of the test program are provided

TABLE 11-4: Data from the experimental program comparing the performance of KCl and NaCl regenerants with both cocurrent and countercurrent regenerant flow

Softener	Regenerant Type	Regenerant Dosage (lbs/ft³)	Capacity (grains)
Softener A cocurrent	NaCl	4	12,742
Softener A cocurrent	NaCl	6	16,140
Softener A cocurrent	NaCl	8	18,417
Softener A cocurrent	NaCl	15	23,459
Softener A cocurrent	KCl	4	13,634
Softener A cocurrent	KCl	6	17,400
Softener A cocurrent	KCl	8	19,686
Softener A cocurrent	KCl	15	24,513
Softener B cocurrent	NaCl	4	13,470
Softener B cocurrent	NaCl	6	15,460
Softener B cocurrent	NaCl	8	17,832
Softener B cocurrent	NaCl	15	21,743
Softener B cocurrent	KCl	4	13,198
Softener B cocurrent	KCl	6	17,108
Softener B cocurrent	KCl	8	19,045
Softener B cocurrent	KCl	15	21,681
Countercurrent	NaCl	4	13,933
Countercurrent	NaCl	6	17,384
Countercurrent	NaCl	8	19,240
Countercurrent	NaCl	15	25,396
Countercurrent	KCl	4	14,939
Countercurrent	KCl	6	18,892
Countercurrent	KCl	8	20,683
Countercurrent	KCl	15	25,655

Note that the softener capacity is defined, in this case, as the total softener capacity (grains) rather than the grains/ft³/pound.

in Table 11-4 and illustrated in Figure 11-5. In Table 11-4 and Figure 11-5 the two cocurrent units are differentiated as softeners A and B.

We need to be careful about interpretation of the data shown in Table 11-4. The goal of this experiment was not to compare performance of cocurrent versus countercurrent water softeners. Rather, the goal was to determine whether the fundamental conclusion from Chapters 7 and 8—that greater efficiency allows KCl to be substituted pound for pound for NaCl with only a modest increase in dosage at the lowest regenerant levels—is applicable to co-current softeners. Consequently, the amount of ion exchange resin in each of the three softeners was not measured precisely, and no comparisons should be made from one softener to the next. Rather, the data in Table 11-4 and Figure 11-5 can be used to check for similarity in the performance of KCl *relative to*

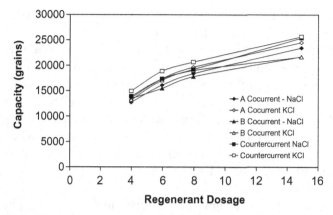

FIGURE 11-5 Total softener capacity (grains) as a function of regenerant dosage (pounds) for KCl and NaCl for several different softeners having either cocurrent or countercurrent regenerant flow. The labels "A" and "B" denote two different cocurrent softeners supplied by two different manufacturers.

the performance of NaCl in the two types of softeners. We can see by inspection of Figure 11-5 that there are no significant differences in the performance of KCl vs. NaCl with both cocurrent and countercurrent softeners. We therefore conclude that the conclusions reached in Chapters 7 and 8 are equally valid for both cocurrent and countercurrent softeners.

PORTABLE EXCHANGE TANKS

Many corporations now offer portable exchange tank (PET) service, in which customers do not own softeners that regenerate automatically. Rather, the company supplies a ready-to-go ion exchange tank. When the resin requires regeneration the company swaps the spent tank for a fresh unit, ready to be put online. The spent unit is then taken to a central processing facility, where it is regenerated. Portable exchange tank service is becoming increasingly popular because of the convenience to the customer, who has no need to purchase and install regenerant on a regular basis. Moreover, portable exchange tank service eliminates the uncontrolled release of spent regenerant waste from each household; rather, the regenerant wastes are produced in a central location, which makes environmentally responsible disposal more practical.

The regeneration of portable exchange tanks with sodium chloride is under a great deal of pressure from government and environmentalists to decrease the sodium and chloride content of the resulting waste brine. The research results provided in Chapters 7 and 8 show that potassium chloride can be substituted for sodium chloride. The use of KCl regenerant eliminates the discharge of sodium to the environment and reduces the amount of chloride waste. Some companies providing portable exchange tank service have switched to KCl regenerant. More would do so except that the cost of KCl is significantly higher than NaCl. Recent research was therefore done to determine whether recycling KCl waste regenerant could be useful as a way of improving efficiency so that the cost of KCl regenerant was more in line with NaCl regenerant.

An experiment was designed to determine the potential benefits of recycling waste KCl regenerant. A set of five laboratory ion exchange columns were assembled in series. Each column was 2 inches in diameter and 36 inches high and contained 1260 grams of commercial water softening ion exchange resin. The series of five ion exchange columns were put through a service cycle until the effluent reached 1 grain/USG hardness. The ion exchange columns were configured so that regenerant could then be routed through one column only, or it could be routed through all five columns in series. Two different types of regeneration were then done:

Single pass, in which a single column was regenerated and then put back into service so that the capacity could be determined.

Multipass, in which the spent regenerant from the first column was routed through the second, third, fourth, and then fifth column.

Once regeneration was completed, the series of five columns was then put back into service. The capacity of the system was then determined by measuring the volume of water treated until the effluent again reached a hardness of 1 grain/USG. Each capacity measurement experiment was repeated several times, and the average capacities are reported in Table 11-5. As with the data discussed in Chapters 7 and 8, it is insightful to calculate and report the capacity per pound of regenerant, as well as the percentage of KCl used, and unused in the actual softening reaction; these results are also provided in Table 11-5.

The data presented in Table 11-5 are illustrated in Figures 11-6 and 11-7. These data show a substantially higher capacity for the multipass configuration. The percentage increase in capacity can be calculated, for example, with 4 pounds of KCl regenerant:

TABLE 11-5: Capacity and percentage of KCl regenerant used and unused (for softening) in a set of single-pass and multipass experiments

Regenerant Dosage (lb)	Single- or Multipass	Capacity (Kgr/ft³)	Capacity (Kgr/ft³/lb)	Percentage Used	Percentage Unused
4	Single	14,200	3,550	75.5	24.5
4	Multi	16,600	4,150	88.3	11.8
6	Single	19,400	3,233	68.8	31.2
6	Multi	23,600	3,933	83.7	16.3
8	Single	22,100	2,763	58.8	41.3
8	Multi	28,500	3,563	75.8	24.3
10	Single	24,300	2,430	51.7	48.3
10	Multi	32,800	3,280	69.8	30.2
12	Single	27,800	2,317	49.3	50.7
12	Multi	35,900	2,992	63.7	36.3
15	Single	29,400	1,960	41.7	58.3

In the single-pass experiment, the waste KCl regenerant from the first column was discarded. In the multipass experiments, the KCl regenerant from the first column was passed through columns 2–5.

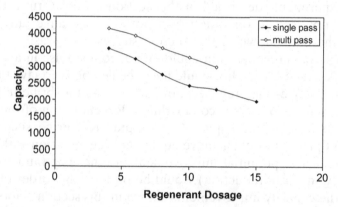

FIGURE 11-6 Capacity per pound of regenerant (Kgrains/ft³/lb) as a function of KCl regenerant dosage (pounds) for single-pass and multipass softening experiments. In the single-pass experiment, the waste KCl regenerant from the first column was discarded. In the multipass experiments, the KCl regenerant from the first column was passed through columns 2–5.

$$\text{Percent Increase} = \left(\frac{4150 - 3550}{3550} \right) \times 100\% = 16.9\%$$

Examination of Table 11-5 shows that such a percent increase in capacity is fairly typical, regardless of the regenerant dosage. Figure 11-7 illustrates the reason for the increase in capacity—the multipass experiment provides higher efficiency because the waste regenerant from the

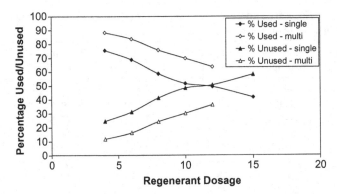

FIGURE 11-7 Percentage of used (for softening) and unused regenerant as a function of KCl regenerant dosage (pounds) for single-pass and multipass softening experiments.

first column remains rich enough in KCl that the second (and subsequent) columns can be regenerated to some extent.

The experiments described in this section do not form a basis for direct comparison to operation of a portable exchange tank facility. Rather, the data shown in this section suggest that waste KCl brine from single-pass regeneration could be put to good use in a softening process. A large PET facility would have the flexibility to save the waste brine from single-stage regeneration, and to use this brine in the first step of regeneration in a second column. Regeneration of the second column could then be completed with a pure KCl brine; such a countercurrent process would achieve equivalent regeneration with the use of less KCl. The potential improvement in efficiency (and therefore, ultimately, in waste reduction) would be on the same order of magnitude as the capacity improvements shown in this section—about 15%.

RESEARCH INTO ALTERNATE REGENERANTS

The search for cost-effective, safe, and environmentally friendly regenerants for water softening was introduced in Chapter 7. As discussed in that chapter, numerous alternatives to NaCl have been investigated because of possible negative impacts of widespread sodium chloride usage in water softeners on human health and the environment. To date, KCl has become the only widely used alternative regenerant, with performance and advantages as discussed in detail in this text. Seawater has also had some limited success as an alternate regenerant, but its use is limited by geography and efficiency.

Research continues into the evaluation of alternative regenerants. Recent studies have been performed on potassium acetate ($KC_2H_3O_2$), which is very soluble in water and nontoxic.

A comparison was made between the capacity of a strong-acid cation exchange resin when regenerated with KCl as opposed to potassium acetate. Details of the experiment were as follows.

— One cubic foot of the resin was regenerated at the 6-pound level by treating with 11,340 grams of a 24% (wt.) solution of KCl. The capacity of the resin at 6 pounds KCl (3.15 pounds of K^+) was then measured and found to be 25.0 Kgrains/ft^3.

— The same cubic foot of resin was then regenerated with 7.9 pounds of potassium acetate, which contains the same quantity of K^+ (3.15 pounds). The capacity of the resin with potassium acetate regeneration was then measured and found to be 26.8 Kgrains/ft^3.

The research thus indicates that, with respect to capacity only, potassium acetate has the potential as a water softener regenerant. The health and safety of using this compound as a regenerant would require further research.

Another alternative to NaCl regenerants has recently been patented (Kunin et al., 2002). In this patent the authors disclose several viable reagents that can regenerate softeners efficiently with no chloride in the regenerant waste.

In the Kunin et al. patent, chloride is eliminated from the regeneration waste by replacing NaCl or KCl with a mixture of three types of chemicals:

— A highly water soluble, nontoxic potassium salt that can be either potassium acetate or potassium formate. The potassium salt provides the K^+ that is able to strip calcium and magnesium from the resin during regeneration.

— A surfactant, designed to retain oils and greases (from upstream pumps) or other organic contaminants so that they do not foul the resin. The surfactant must be anionic, since cationic or nonionic surfactants would bind to (and foul) the negatively charged sulfonic acid functional group on a strong acid cation exchanger. In addition, the surfactant must be safe for human consumption and soluble in the regenerant brine and must not react with hardness ions (which would lead to formation of a precipitate). The preferred surfactant was disclosed as octyl phenol ethoxylate.

— A chelating (sequestering) compound, designed to bind iron and other metal ions and thereby prevent fouling of the ion exchange resin. The preferred embodiment disclosed was citric acid, but other chelating agents were also discussed, including ethylene-diamine tetraacetic acid (EDTA), amino phosphonates, and gluconic acid.

The patent by Kunin et al. describes the use of a liquid regenerant, that is, the potassium acetate or formate is sold as a concentrated solution in water. The use of a liquid regenerant allows for uniform distribution of the surfactant and chelating components, which provides effective protection from fouling.

The results presented in this section are not comprehensive. Rather, they are just a sampling of the ongoing research designed to find improved regenerants able to soften water with reduced environmental impact. Further research into alternate regenerants is ongoing.

REFERENCES

Henke L. Softeners remove iron from groundwater. Water Conditioning Purification 1995; May: 68–69.

Kunin R et al. Non-chloride containing regenerant composition for a strong acid cation exchange resin of a water softener. United States Patent 6340712. 2002 January 22.

STANDARD TEST PROTOCOL FOR COMPARISON OF KCl AND NaCl REGENERANTS

To facilitate meaningful comparisons between the performance of KCl and NaCl as regenerants, it was necessary to develop a standardized test protocol. The results provided in this appendix describe the test protocol, which was developed in consultation with the Water Quality Association (WQA) as well as the National Sanitation Foundation (NSF).

PROTOCOL DEVELOPMENT

A set of goals, or objectives, were established for the test protocol, which are described below. In addition, comments have been added to explain what the motivation was for the objective. The following objectives were established for the protocol:

1. To determine the weight of KCl:NaCl per cubic foot of resin required to displace an equal amount of hardness from the softener resin during regeneration.

   This information can be used in advising consumers on any necessary changes in water softener settings when changing from use of NaCl to use of KCl for regeneration.

Water Softening with Potassium Chloride: Process, Health, and Environmental Benefits, by William Wist, Jay H. Lehr, and Rod McEachern
Copyright © 2009 by John Wiley & Sons, Inc.

2. a) To determine the relationship between the weight of KCl or NaCl regenerant per cubic foot of resin and the percentage of the resin capacity regenerated.

    b) To determine the relationship between the weight of KCl or NaCl regenerant per cubic foot of resin and the percentage of regenerant losses to the sewer.

    c) To determine the relationship between the percentage of the resin capacity regenerated and the percentage of regenerant losses to the sewer for both KCl and NaCl regenerants.

    This information can be used in illustrating how the uses of higher regenerant levels provides a small increase in the length of the service cycle but a large increase in losses to the regenerant effluent.

3. To determine the effect of the use of KCl or NaCl as a regenerant on the K and Na content of the softener discharge.

    This information can be used in illustrating the quality of the water produced during the service cycle.

4. To determine the effect of the use of KCl or NaCl as a regenerant on the Ca and Mg content of the softener discharge.

    This information can be used in illustrating the quality of the water produced during the service cycle.

The protocol was developed to provide the following data which would be processed to achieve the objectives outlined above:

1. Determine the weight of KCl:NaCl per cubic foot of resin required to displace an equal amount of hardness from the softener resin during regeneration.

    a) Regenerate the resin with a known weight of regenerant.

    b) Displace the regenerant from the resin with raw water.

    c) Feed raw water through the resin until the softener effluent contains 1 grain per gallon of hardness. The total hardness picked up by the softener resin equals the amount of raw water passed though the resin, multiplied by the raw water hardness.

Four replications are performed at each of three regenerant levels for both KCl and NaCl. The test thus provides twelve KCl readings which can be compared with twelve NaCl readings.

2. a) Determine the relationship between the weight of KCl or NaCl regenerant per cubic foot of resin and the percentage of the resin capacity regenerated.

    The data produced in (1) above are analyzed to determine this relationship.

    Four KCl and four NaCl data points are obtained at each of three regenerant levels.

  b) Determine the relationship between the weight of KCl or NaCl regenerant per cubic foot of resin and the percentage of regenerant losses to the sewer.

    The data produced in (1) above are analyzed to determine this relationship.

    Four KCl and four NaCl data points are obtained at each of three regenerant levels.

  c) Determine the relationship between the percentage of the resin capacity regenerated and the percentage of regenerant losses to the sewer for both KCl and NaCl regenerants.

    The data from 2(a) and 2(b) above are used to determine this relationship.

3. Determine the effect of the use of KCl or NaCl as a regenerant on the K and Na content of the softener discharge.

    For the first three of the four replications performed at each regenerant level in (1) above, determine the raw water volume required to obtain one grain per US gallon hardness in the softener effluents. This is 100% capacity for the regenerant level used. For the fourth replication, collect samples at 20, 75, 90, and 100 percent of the capacity established in the initial three replications.

    Analyze the samples collected at 20, 75, 90, and 100 percent of capacity for Ca, Mg, K, and Na.

    One replication is performed at each of three regenerant levels for both KCl and NaCl.

    The above set of analysis provides four points along a curve for each of three regenerant levels for both KCl and NaCl.

4. Determine the effect of the use of KCl or NaCl as a regenerant on the Ca and Mg content of the softener discharge.

    The data produced in (3) above are used to determine this effect.

    This provides four points along a curve for each of three regenerant levels for both KCl and NaCl.

A test protocol that included two raw waters was initially considered. One raw water contained 30 grains hardness per US gallon, with the hardness consisting of calcium and magnesium in a ratio of 2:1. The second raw water also contained 30 grains hardness per US gallon, but the hardness was present in a calcium:magnesium ratio of 1:2. Test work carried out at the PotashCorp pilot plant during the period of protocol development determined that the two raw waters gave similar performance. Subsequent to this test work, the raw water having a calcium:magnesium hardness ratio of 1:2 was removed from the protocol.

STANDARD TEST PROTOCOL

The testing described under this protocol was designed to directly compare the efficiency of three different levels of potassium chloride (KCl) versus sodium chloride (NaCl) in regenerating household cation exchange media and to compare the quantity and quality of the softened water produced with each regenerant.

Apparatus

Five ECO Water Systems Model 2650 water softeners were purchased. The resin was removed from all five softeners and transferred to a large plastic container and mixed well. The softener tanks were washed to remove any remaining resin.

Each softener tank was tared on a balance (±1 gram) and resin was added until a weight of 24,267 grams (1 ft^3) was added. The four softeners were randomly selected for the four participants and shipped to the respective sites immediately.

Each site installed and operated the ECO Water Systems Model 2650 countercurrent cation exchange water softener containing 1 ft^3 resin (24.267 kg) as received. The units were installed as shown in Figure A1-1. The units were connected to the water supply in accordance with the standard installation instructions issued by PotashCorp.

The two regenerants, KCl WSM 0.2 from PCS Cory and NaCl Aqua-Magic Crystal Plus from Sifto Salt, were also shipped to the participants.

Test Water and Regenerants

If available, a potable water supply can be used, provided it has the following characteristics:

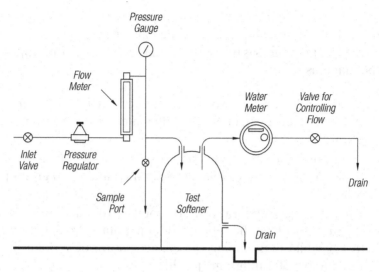

FIGURE A1-1 Configuration of equipment in the standard water test protocol apparatus.

Hardness (as $CaCO_3$)	$20\,Ca:10\,Mg \pm 1.0$ grains/US gallon
Iron	$\leq 0.1\,mg$ / liter
Sodium/potassium salts	$1:1$ weight ratio, each less than $50\,ppm$
TDS	$\leq 1000\,mg$ / liter
Turbidity	<1.0 NTU
pH	7.5 ± 0.5
Temperature	$20 \pm 2\,°C$

If the local potable water supply does not meet the above characteristics, then the test water can be synthesized in accordance with the specifications given in the following section.

Each batch of raw water made up will be analyzed to ensure that it has the desired chemical content before use.

Regenerant salts are to be shipped to the participants in the test program from a single source as arranged by PotashCorp. The KCl and NaCl regenerant solutions are to be prepared as 24% (w/w) to provide unsaturated solutions that can accommodate temperature fluctuations without precipitation.

Synthesis of Test Water

If necessary, test water can be prepared from the existing potable water supply with the following procedure.

Adjust the raw water to 20 ± 1 grains calcium:10 ± 1 grains magnesium per US gallon using the chloride salt of calcium and/or magnesium. The total hardness is to be 30 ± 1 grains per US gallon. Adjustments can be made as follows:

a) Calcium adjustment: Add 95.0 grams of $CaCl_2 \cdot 2H_2O$ per 1000 US gallons water to achieve an adjustment of 1.0 grains per US gallon.

b) Magnesium adjustment: Add 131.4 grams $MgCl_2 \cdot 6H_2O$ per 1000 US gallons water to achieve an increase of 1.0 grains per US gallons.

c) Sodium:potassium ratio adjustment: Add KCl or NaCl as required to adjust the ratio of sodium:potassium in the raw water to 1:1 while maintaining the concentration of sodium or potassium below 50 milligrams per liter.

Add the required amount of chemicals to produce the raw water specified in the "Test Water and Regenerants" section. Mix thoroughly to ensure homogeneity. Analyze each batch before use to ensure the water meets the standards specified in the "Test Water and Regenerants" section.

Analytical Methods

Perform all water sample analyses for Na, K, Mg, and Ca by atomic absorption, inductively coupled plasma, or ion chromatography, using the methodology described in "Standard Methods for the Examination of Water and Waste Water" (Clescerl et al., 1999). Supplementary analysis of water samples by the EDTA volumetric method may, at the discretion of the testing laboratory, also be performed; however, such analysis will not be reported.

Analytical requirements for each test run are given in Table A1-1, and the total analytical requirements for the test protocol are summarized in Table A1-2.

Test Procedure

1. All test runs utilizing NaCl will be completed before using KCl to regenerate the softener. Test conditions for each of the runs in the protocol are summarized in Table A1-3, and described in detail below.

TABLE A1-1: Analytical requirements for each test run of the standardized test protocol

Test Run	Regeneration Cycle		Service Cycle		Additional Effluent Analysis
	Influent	Effluent	Influent	Effluent	
1-1	none	none	none	T.H. 1 gpg EDTA	
1-2	none	none	none	"	
1-3	none	none	none	"	
2-1	Composite of Cuts Ca, Mg, K, Na, Cl	Total Sample Ca, Mg, K, Na, Cl	Composite of Cuts Ca, Mg, K, Na, Cl	End pt. deter. T.H. 1 gpg EDTA	
2-2	"	"	"	"	
2-3	"	"	"	"	
2-4	"	"	"	"	20,75,90,100 Ca, Mg, Ka, Na
3-1	Composite of Cuts Ca, Mg, K, Na, Cl	Total Sample Ca, Mg, K, Na, Cl	Composite of Cuts Ca, Mg, K, Na, Cl	End pt. deter. T.H. 1 gpg EDTA	
3-2	"	"	"	"	
3-3	"	"	"	"	
3-4	"	"	"	"	20,75,90,100 Ca, Mg, Ka, Na
4-1	Composite of Cuts Ca, Mg, K, Na, Cl	Total Sample Ca, Mg, K, Na, Cl	Composite of Cuts Ca, Mg, K, Na, Cl	End pt. deter. T.H. 1 gpg EDTA	
4-2	"	"	"	"	
4-3	"	"	"	"	
4-4	"	"	"	"	20,75,90,100 Ca, Mg, Ka, Na
5-1	none	none	none	T.H. 1 gpg EDTA	
5-2	none	none	none	"	
5-3	none	none	none	"	
6-1	Composite of Cuts Ca, Mg, K, Na, Cl	Total Sample Ca, Mg, K, Na, Cl	Composite of Cuts Ca, Mg, K, Na, Cl	End pt. deter. T.H. 1 gpg EDTA	
6-2	"	"	"	"	
6-3	"	"	"	"	
6-4	"	"	"	"	20,75,90,100 Ca, Mg, Ka, Na
7-1	Composite of Cuts Ca, Mg, K, Na, Cl	Total Sample Ca, Mg, K, Na, Cl	Composite of Cuts Ca, Mg, K, Na, Cl	End pt. deter. T.H. 1 gpg EDTA	

TABLE A1-1: *Continued*

Test Run	Regeneration Cycle		Service Cycle		Additional Effluent Analysis
	Influent	Effluent	Influent	Effluent	
7-2	"	"	"	"	
7-3	"	"	"	"	
7-4	"	"	"	"	20,75,90,100 Ca, Mg, Ka, Na
8-1	Composite of Cuts Ca, Mg, K, Na, Cl	Total Sample Ca, Mg, K, Na, Cl	Composite of Cuts Ca, Mg, K, Na, Cl	End pt. deter. T.H. 1 gpg EDTA	
8-2	"	"	"	"	
8-3	"	"	"	"	
8-4	"	"	"	"	20,75,90,100 Ca, Mg, Ka, Na

TABLE A1-2: Total analytical requirements by element for test runs in the standardized test protocol

Element	Regeneration Cycle		Service Cycle		Total
	Influent	Effluent	Influent	Effluent	
Ca	24	24	24	24	96
Mg	24	24	24	24	96
K	24	24	24	24	96
Na	24	24	24	24	96
Cl	24	24	none	none	48

TABLE A1-3: Summary of softener test runs in the standardized test protocol

Test	Number of Replications	Regenerant	Pounds	Purpose
1-1	1	NaCl	25	Resin conditioning
1-2, 1-3	2	NaCl	15	Resin conditioning
2-1 to 2-4	4	NaCl	4	Regenerant evaluation
3-1 to 3-4	4	NaCl	8	Regenerant evaluation
4-1 to 4-4	4	NaCl	15	Regenerant evaluation
5-1	1	KCl	25	Resin conditioning
5-2, 5-3	2	KCl	15	Resin conditioning
6-1 to 6-4	4	KCl	4	Regenerant evaluation
7-1 to 7-4	4	KCl	8	Regenerant evaluation
8-1 to 8-4	4	KCl	15	Regenerant evaluation

2. Precondition the softener before the actual series of test runs with three complete softening runs: one at 25 lbs and two at 15 lbs of the regenerant to be used.

3. Conduct the actual test runs at each of three salt levels: 4, 8, and 15 lbs. There is to be a minimum of four softening runs at each level of regeneration.

   a) Adjust the water pressure regulator to establish a constant operating pressure of 35 ± 5 psi.

   b) Set the softener salt dosage and regenerate the softener per manufacturer's instructions. Collect a composite sample that contains both the regeneration cycle effluent and the displacement wash used to remove the regenerant from the resin before beginning the service cycle. Weigh the composite sample (or determine weight from volume and specific gravity) and analyze for Ca, Mg, K, Na, and chloride.

   c) Record the initial water meter reading. Establish the softener service flow rate at 50% of the manufacturer's rated service flow. Rated service flow for the Eco Water Systems model 2650 water softener is 8.0 US gallons per minute; therefore, service flow rate should be set at 4.0 US gpm.

   d) Begin the test and collect effluent water samples for analysis as follows:

      i) For the first three runs of a series, collect samples as required to accurately determine the endpoint of 1.0 grains/US gallon.

      ii) On the fourth run, collect samples at 20, 75, and 90 percent of the capacity established in the first three runs. Determine the endpoint of the fourth run in the normal way and collect a sample at 100% capacity.

      iii) The four samples will be analyzed for Ca, Mg, K, and Na.

4. The grains of hardness removed shall be the gallons of softened water produced with a hardness level <1 grains/US gallon as $CaCO_3$, multiplied by the average hardness of the raw water.

   a) To determine the hardness of the influent water, a percentage of influent water shall be collected during the test run and the composite sample analyzed for hardness by acceptable water hardness test methods under this protocol. Measure and record the TDS at the start and end of each test run.

5. The overall average capacity (grains removed) at each salt level shall be based on the mean of a minimum of three successive

test runs, which deviate <10% from the mean of all four test runs.

6. Each test run will be carried out as a continuous run. A test run will not be partially completed one day and completed the next day.

Test Records and Calculations

Data from each test run shall be recorded as follows:

1. Record ambient laboratory temperature.
2. Record influent water temperature.
3. Record hardness of influent composite water sample.
4. Record amount of salt in pounds for each regeneration.
5. Record the total gallons of soft water produced.

Results

Record the salt efficiency, expressed in grains of exchange per pound of salt, at each salt setting, based on the mean grains of hardness removed as established in the "Test Procedure" section (above) item 4.

REFERENCE

Clescerl LS, Greenberg AE, Eaton AD, editors. *Standard Methods for the Examination of Water and Wastewater*. American Public Health Association; 20th edition; January 1999.

APPENDIX 2

LABORATORY DATA OBTAINED WITH THE STANDARD TEST PROTOCOL

To obtain a valid, credible comparison between the regeneration performance of KCl and NaCl, the standard test protocol described in Appendix 1 was used independently by four different laboratories. The detailed test results obtained by each laboratory are presented in this appendix, while the implications of the data are discussed in detail in Chapter 8.

Participating laboratories in the study were: National Sanitation Foundation, PotashCorp, Sifto Salt, and the Water Quality Association. In the data to follow, individual laboratory names are not provided; rather, individual labs are identified only as "A", "B", "C", and "D."

Each laboratory followed the test protocol described in detail in Appendix 1. Data and calculated values obtained from each test included:

— Type of regenerant used (KCl or NaCl)
— Regenerant dosage in pounds
— US gallons of influent treated until the effluent hardness exceeded 1 grain per gallon (as $CaCO_3$).

Water Softening with Potassium Chloride: Process, Health, and Environmental Benefits, by William Wist, Jay H. Lehr, and Rod McEachern
Copyright © 2009 by John Wiley & Sons, Inc.

— Grains hardness removed per cubic foot of resin. In each case the volume of resin was one cubic foot, so this parameter was calculated simply as:

$$\text{Grains}/\text{ft}^3 = (\text{Gallons treated}) \times (\text{Influent hardness})$$

— Grains hardness removed per cubic foot of resin per pound of regenerant.
— Pounds of regenerant used. This parameter was calculated by comparison of the grains of hardness removed per cubic foot of resin per pound of regenerant, with the theoretical values (calculated in Chapter 8). For example, we know that NaCl has a theoretical regeneration capacity of 5998 grains per pound. Therefore, a test run that achieved 4500 grains hardness removal per cubic foot of resin per pound of regenerant actually used:

$$\frac{4500}{5998} \times 100\% = 75.03\%$$

of the NaCl for softening. In the present example, 4 pounds of regenerant were used, so the pounds of NaCl actually used was then calculated:

$$4.0 \text{ pounds} \times 0.7503 = 3.00 \text{ pounds}$$

Similarly, pounds of KCl used was calculated by comparison to the theoretical capacity of 4702 grains hardness removed per pound of regenerant, as described in Chapter 8.
— Pounds of regenerant unused. Calculated as the difference between the pounds of regenerant in the test and the pounds used for hardness removal (calculated above).
— Percentage of regenerant used. Calculated as the ratio of the regenerant used (for hardness removal) relative to the total pounds of regenerant used in the test, and reported as a percentage.

The test protocol, as described in Appendix 1, included steps for resin conditioning. The conditioning steps included tests 1-1, 1-2, 1-3, 5-1, 5-2, and 5-3 as described in Table A1-3 of Appendix 1. Performance of the regenerant was not calculated based on these conditioning steps, so raw data for these steps of the process are not included in the following tables.

TABLE A2-1: Results of KCl/NaCl comparison study, as per the Standard Test Protocol (Appendix 1) for Laboratory "A" with NaCl as regenerant

Test	Type of Regenerant	Pounds Regenerant	US Gallons	Influent Hardness (gpg as CaCO₃)	Grains Per ft³	Grains Per ft³/lb	Pounds Regenerant Used	Pounds Regenerant Unused	Percent Regenerant Unused
2-1	NaCl	4	600	30.0	18,000	4,500	3.00	1.00	24.94
2-2	NaCl	4	580	29.9	17,342	4,336	2.89	1.11	27.68
2-3	NaCl	4	643	31.1	19,997	4,999	3.34	0.66	16.61
2-4	NaCl	4	560	30.9	17,304	4,326	2.89	1.11	27.84
Ave.	**NaCl**	**4**	**596**	**30.5**	**18,161**	**4,540**	**3.03**	**0.97**	**24.27**
3-1	NaCl	8	860	29.7	25,542	3,193	4.26	3.74	46.74
3-2	NaCl	8	930	29.0	26,970	3,371	4.50	3.50	43.77
3-3	NaCl	8	860	30.2	25,972	3,247	4.33	3.67	45.85
3-4	NaCl	8	870	29.2	25,404	3,176	4.24	3.76	47.03
Ave.	**NaCl**	**8**	**880**	**29.5**	**25,972**	**3,247**	**4.33**	**3.67**	**45.85**
4-1	NaCl	15	1,077	30.2	32,525	2,168	5.43	9.57	63.83
4-2	NaCl	15	1,145	29.9	34,236	2,282	5.71	9.29	61.93
4-3	NaCl	15	1,150	30.6	35,190	2,346	5.87	9.13	60.87
4-4	NaCl	15	1,162	31.0	36,022	2,401	6.01	8.99	59.94
Ave.	**NaCl**	**15**	**1,134**	**30.4**	**34,493**	**2,300**	**5.75**	**9.25**	**61.64**

TABLE A2-2: Results of KCl/NaCl comparison study, as per the Standard Test Protocol (Appendix 1) for Laboratory "A" with KCl as regenerant

Test	Type of Regenerant	Pounds Regenerant	US Gallons	Influent Hardness (gpg as $CaCO_3$)	Grains Per ft^3	Grains Per ft^3/lb	Pounds Regenerant Used	Pounds Regenerant Unused	Percent Regenerant Unused
6-1	KCl	4	560	29.1	16,296	4,074	3.47	0.53	13.31
6-2	KCl	4	500	29.7	14,850	3,713	3.16	0.84	21.00
6-3	KCl	4	480	29.2	14,016	3,504	2.98	1.02	25.43
6-4	KCl	4	465	30.3	14,090	3,522	3.00	1.00	25.04
Ave.	**KCl**	**4**	**501**	**29.6**	**14,813**	**3,703**	**3.15**	**0.85**	**21.20**
7-1	KCl	8	900	30.0	27,000	3,375	5.75	2.25	28.18
7-2	KCl	8	900	29.9	26,910	3,364	5.73	2.27	28.42
7-3	KCl	8	940	30.2	28,388	3,549	6.04	1.96	24.49
7-4	KCl	8	920	31.0	28,520	3,565	6.07	1.93	24.14
Ave.	**KCl**	**8**	**915**	**30.3**	**27,705**	**3,463**	**5.90**	**2.10**	**26.31**
8-1	KCl	15	1,080	29.0	31,320	2,088	6.66	8.34	55.57
8-2	KCl	15	1,080	30.2	32,616	2,174	6.94	8.06	53.73
8-3	KCl	15	1,100	29.4	32,340	2,156	6.88	8.12	54.12
8-4	KCl	15	1,080	30.4	32,832	2,189	6.99	8.01	53.42
Ave.	**KCl**	**15**	**1,085**	**29.8**	**32,277**	**2,152**	**6.87**	**8.13**	**54.21**

TABLE A2-3: Results of KCl/NaCl comparison study, as per the Standard Test Protocol (Appendix 1) for Laboratory "B" with NaCl as regenerant

Test	Type of Regenerant	Pounds Regenerant	US Gallons	Influent Hardness (gpg as CaCO₃)	Grains Per ft³	Grains Per ft³/lb	Pounds Regenerant Used	Pounds Regenerant Unused	Percent Regenerant Unused
2-1	NaCl	4	366	30.5	11,163	2,791	1.86	2.14	53.45
2-2	NaCl	4	374	32.3	12,080	3,020	2.01	1.99	49.63
2-3	NaCl	4	394	31.8	12,529	3,132	2.09	1.91	47.75
2-4	NaCl	4	396	31.8	12,593	3,148	2.10	1.90	47.49
Ave.	**NaCl**	**4**	**383**	**31.6**	**12,091**	**3,023**	**2.02**	**1.98**	**49.58**
3-1	NaCl	8	715	31.5	22,523	2,815	3.76	4.24	53.04
3-2	NaCl	8	727	31.1	22,610	2,826	3.77	4.23	52.86
3-3	NaCl	8	751	31.1	23,356	2,920	3.90	4.10	51.30
3-4	NaCl	8.4	798	30.3	24,179	2,879	4.03	4.37	51.99
Ave.	**NaCl**	**8.1**	**748**	**31.0**	**23,167**	**2,860**	**3.86**	**4.24**	**52.30**
4-1	NaCl	15	958	30.9	29,602	1,973	4.94	10.06	67.08
4-2	NaCl	15	948	31.7	30,052	2,003	5.01	9.99	66.58
4-3	NaCl	15	1,000	30.5	30,500	2,033	5.09	9.91	66.08
4-4	NaCl	15	946	32.0	30,272	2,018	5.05	9.95	66.34
Ave.	**NaCl**	**15**	**963**	**31.3**	**30,106**	**2,007**	**5.02**	**9.98**	**66.52**

TABLE A2-4: Results of KCl/NaCl comparison study, as per the Standard Test Protocol (Appendix 1) for Laboratory "B" with KCl as regenerant

Test	Type of Regenerant	Pounds Regenerant	US Gallons	Influent Hardness (gpg as CaCO$_3$)	Grains Per ft^3	Grains Per ft^3/lb	Pounds Regenerant Used	Pounds Regenerant Unused	Percent Regenerant Unused
6-1	KCl	4	328	32.6	10,693	2,673	2.28	1.72	43.11
6-2	KCl	4	325	31.7	10,303	2,576	2.19	1.81	45.19
6-3	KCl	4	322	33.9	10,916	2,729	2.32	1.68	41.93
6-4	KCl	4	350	31.5	11,025	2,756	2.35	1.65	41.35
Ave.	**KCl**	**4**	**331**	**32.4**	**10,734**	**2,684**	**2.28**	**1.72**	**42.89**
7-1	KCl	8	710	31.6	22,436	2,805	4.77	3.23	40.32
7-2	KCl	8	712	32.1	22,855	2,857	4.86	3.14	39.21
7-3	KCl	8	730	32.3	23,579	2,947	5.02	2.98	37.28
7-4	KCl	8	684	31.8	21,751	2,719	4.63	3.37	42.14
Ave.	**KCl**	**8**	**709**	**32.0**	**22,655**	**2,832**	**4.82**	**3.18**	**39.74**
8-1	KCl	15	926	32.9	30,465	2,031	6.48	8.52	56.78
8-2	KCl	15	959	31.9	30,592	2,039	6.51	8.49	56.60
8-3	KCl	15	958	32.5	31,135	2,076	6.63	8.37	55.83
8-4	KCl	15	937	32.3	30,265	2,018	6.44	8.56	57.06
Ave.	**KCl**	**15**	**945**	**32.4**	**30,614**	**2,041**	**6.51**	**8.49**	**56.57**

TABLE A2-5: Results of KCl/NaCl comparison study, as per the Standard Test Protocol (Appendix 1) for Laboratory "C" with NaCl as regenerant

Test	Type of Regenerant	Pounds Regenerant	US Gallons	Influent Hardness (gpg as CaCO$_3$)	Grains Per ft^3	Grains Per ft^3/lb	Pounds Regenerant Used	Pounds Regenerant Unused	Percent Regenerant Unused
2-1	NaCl	4	527	29.6	15,600	3,900	2.60	1.40	34.95
2-2	NaCl	4	545	29.0	15,800	3,950	2.64	1.36	34.11
2-3	NaCl	4	541	29.2	15,800	3,950	2.64	1.36	34.11
2-4	NaCl	4	527	29.6	15,600	3,900	2.60	1.40	34.95
Ave.	**NaCl**	**4**	**535**	**29.4**	**15,700**	**3,925**	**2.62**	**1.38**	**34.53**
3-1	NaCl	8	829	28.7	23,800	2,975	3.97	4.03	50.38
3-2	NaCl	8	722	32.7	23,600	2,950	3.94	4.06	50.79
3-3	NaCl	8	810	29.5	23,900	2,988	3.99	4.01	50.17
3-4	NaCl	8	799	29.3	23,400	2,925	3.90	4.10	51.21
Ave.	**NaCl**	**8**	**790**	**30.1**	**23,675**	**2,959**	**3.95**	**4.05**	**50.64**
4-1	NaCl	15	1,083	30.3	32,800	2,187	5.47	9.53	63.53
4-2	NaCl	15	1,079	30.2	32,600	2,173	5.44	9.56	63.75
4-3	NaCl	15	1,101	29.6	32,600	2,173	5.44	9.56	63.75
4-4	NaCl	15	1,125	28.9	32,500	2,167	5.42	9.58	63.86
Ave.	**NaCl**	**15**	**1,097**	**29.8**	**32,625**	**2,175**	**5.44**	**9.56**	**63.72**

TABLE A2-6: Results of KCl/NaCl comparison study, as per the Standard Test Protocol (Appendix 1) for Laboratory "C" with KCl as regenerant

Test	Type of Regenerant	Pounds Regenerant	US Gallons	Influent Hardness (gpg as CaCO₃)	Grains Per ft³	Grains Per ft³/lb	Pounds Regenerant Used	Pounds Regenerant Unused	Percent Regenerant Unused
6-1	KCl	4	471	29.5	13,900	3,475	2.96	1.04	26.05
6-2	KCl	4	459	29.6	13,600	3,400	2.89	1.11	27.65
6-3	KCl	4	455	29.9	13,600	3,400	2.89	1.11	27.65
6-4	KCl	4	488	28.3	13,800	3,450	2.94	1.06	26.58
Ave.	**KCl**	**4**	**468**	**29.3**	**13,725**	**3,431**	**2.92**	**1.08**	**26.98**
7-1	KCl	8	744	29.3	21,800	2,725	4.64	3.36	42.01
7-2	KCl	8	719	29.9	21,500	2,688	4.58	3.42	42.81
7-3	KCl	8	727	29.3	21,300	2,663	4.53	3.47	43.34
7-4	KCl	8	734	29.0	21,300	2,663	4.53	3.47	43.34
Ave.	**KCl**	**8**	**731**	**29.4**	**21,475**	**2,684**	**4.57**	**3.43**	**42.88**
8-1	KCl	15	1,101	29.6	32,600	2,173	6.94	8.06	53.75
8-2	KCl	15	1,106	29.2	32,300	2,153	6.87	8.13	54.18
8-3	KCl	15	1,028	31.9	32,800	2,187	6.98	8.02	53.47
8-4	KCl	15	1,070	30.1	32,200	2,147	6.85	8.15	54.32
Ave.	**KCl**	**15**	**1,076**	**30.2**	**32,475**	**2,165**	**6.91**	**8.09**	**53.93**

TABLE A2-7: Results of KCl/NaCl comparison study, as per the Standard Test Protocol (Appendix 1) for Laboratory "D" with NaCl as regenerant

Test	Type of Regenerant	Pounds Regenerant	US Gallons	Influent Hardness (gpg as CaCO$_3$)	Grains Per ft^3	Grains Per ft^3/lb	Pounds Regenerant Used	Pounds Regenerant Unused	Percent Regenerant Unused
2-1	NaCl	4	657	30.17	19,822	4,955	3.31	0.69	17.25
2-2	NaCl	4	656	30.17	19,792	4,948	3.30	0.70	17.50
2-3	NaCl	4	651	30.17	19,641	4,910	3.28	0.72	18.00
2-4	NaCl	4	646	30.13	19,464	4,866	3.25	0.75	18.75
Ave.	**NaCl**	**4**	**653**	**30.16**	**19,679**	**4,920**	**3.28**	**0.72**	**18.00**
3-1	NaCl	8	985	30.13	29,678	3,710	4.95	3.05	38.13
3-2	NaCl	8	973	30.13	29,316	3,665	4.89	3.11	38.88
3-3	NaCl	8	972	30.13	29,286	3,661	4.88	3.12	39.00
3-4	NaCl	8	990	30.13	29,820	3,729	4.98	3.02	37.75
Ave.	**NaCl**	**8**	**980**	**30.13**	**29,527**	**3,691**	**4.93**	**3.07**	**38.38**
4-1	NaCl	15	1,186	30.11	35,710	2,381	5.96	9.04	60.27
4-2	NaCl	15	1,205	30.11	36,283	2,419	6.05	8.95	59.67
4-3	NaCl	15	1,195	29.85	35,671	2,378	5.95	9.05	60.33
4-4	NaCl	15	1,202	29.85	35,880	2,392	5.98	9.02	60.13
Ave.	**NaCl**	**15**	**1,197**	**29.98**	**35,886**	**2,392**	**5.99**	**9.01**	**60.07**

TABLE A2-8: Results of KCl/NaCl comparison study, as per the Standard Test Protocol (Appendix 1) for Laboratory "D" with KCl as regenerant

Test	Type of Regenerant	Pounds Regenerant	US Gallons	Influent Hardness (gpg as CaCO$_3$)	Grains Per ft^3	Grains Per ft^3/lb	Pounds Regenerant Used	Pounds Regenerant Unused	Percent Regenerant Unused
6-1	KCl	4	639	29.85	19,074	4,769	4.06	-0.06	-1.50
6-2	KCl	4	618	29.85	18,447	4,612	3.93	0.07	1.75
6-3	KCl	4	645	29.94	19,311	4,828	4.11	-0.11	-2.75
6-4	KCl	4	606	29.94	18,144	4,536	3.86	0.14	3.50
Ave.	**KCl**	**4**	**627**	**29.90**	**18,747**	**4,687**	**3.99**	**0.01**	**0.25**
7-1	KCl	8	951	29.94	28,473	3,559	6.06	1.94	24.25
7-2	KCl	8	980	29.94	29,341	3,668	6.24	1.76	22.00
7-3	KCl	8	978	29.94	29,281	3,660	6.23	1.77	22.13
7-4	KCl	8	990	29.94	29,641	3,705	6.31	1.69	21.13
Ave.	**KCl**	**8**	**975**	**29.94**	**29,184**	**3,648**	**6.21**	**1.79**	**22.38**
8-1	KCl	15	1,167	29.94	34,940	2,329	7.44	7.56	50.40
8-2	KCl	15	1,144	30.03	34,354	2,290	7.31	7.69	51.27
8-3	KCl	15	1,167	30.03	35,045	2,336	7.46	7.54	50.27
8-4	KCl	15	1,166	30.03	35,015	2,334	7.45	7.55	50.33
Ave.	**KCl**	**15**	**1,161**	**30.01**	**34,842**	**2,323**	**7.41**	**7.59**	**50.60**

APPENDIX 3

ACCELERATED MUSH TEST

The accelerated mush test described herein was developed to measure the performance quality of compacted water softening salts. The procedure measures the volume of fine particles generated as a result of immersion and vigorous agitation in brine.

APPARATUS

The following apparatus is required to perform the accelerated mush test:

- Laboratory balance
- Wide-mouth 16-ounce plastic jars with lids
- Three half-height sieve frames with screens removed
- Eight half-height test sieves, including a Tyler 4 mesh
- A half-height sieve pan and cover for the nest of sieves
- Tyler 8 mesh sieve
- Ro Tap sieve shaker
- Large plastic funnel

Water Softening with Potassium Chloride: Process, Health, and Environmental Benefits, by William Wist, Jay H. Lehr, and Rod McEachern
Copyright © 2009 by John Wiley & Sons, Inc.

- Imhoff settling cones
- Saturated KCl and NaCl brine

METHOD

Precisely weigh 200 grams of regenerant sample and transfer to one of the plastic jars. Only whole pellets or buttons of the regenerant should be selected for the test.

Pour 200 ml of the appropriate saturated brine (i.e., KCl-saturated brine for KCl regenerant, and NaCl-saturated brine for NaCl regenerant) into the jar. Tightly screw on the tap, and seal the jar with electrical tape to prevent leaking or evaporation of water.

Repeat the weighing and brine addition step for up to three different samples.

Store the sample jars in an undisturbed location. The storage time is typically 24 hours, but time series experiments can also be performed over a longer time span. Record the precise storage time.

Transfer the sample jars to the nest of sieves. The nest should consist of, from top to bottom: the three sieve rings without sieves, the Tyler 4 mesh sieve, seven other half-height sieves, and a half-height pan. Three jars must be placed resting on the 4-mesh sieve to ensure a snug fit. If less than three samples are to be processed at one time, then additional jars (containing the appropriate weight of brine and/or regenerant) can be used.

Ro tap for precisely 5 minutes.

Insert the large plastic funnel into one of the Imhoff cones, and insert the Tyler 8 mesh screen inside the funnel.

Remove one of the sample jars from the Ro tap and pour its contents through the 8-mesh screen into the Imhoff cone. Rinse the sample jar with the appropriate saturated brine until all fines have been transferred to the cone.

Repeat the transfer of material to the other Imhoff cones, if more than one sample is being analyzed.

Allow the fine particulate to settle for 2 hours, occasionally tapping the sides of the cone to encourage uniform packing. Record the volume of fines obtained, and report.

INDEX

Accelerated mush test, 144, 243–244
Activated sludge process, 164–166
Affinity, 75–76, 150
Air bubbles, 19–21
Alcohols, 20
Algae, 166–168
  bloom(s), 167–168, 177
Alternate regenerants, 99–102, 220–222
Amine
  collector, 19–20
  functional group, 57
Anaerobic digestion, 165–166
Analytical methods, 104, 228
Anhydrite, 17, 29
Anion exchange, 45–46, 71–72
Anions, 42, 45–46, 83–84
Atomic
  mass, 79–82
  number, 79–80
  orbital, 79

size, 76
weight(s), 76, 80–82, 146–156

Backwash, 64, 90–92, 94
  in sewage, 163–164
Bacteria, 162–166, 169
  inhibition, 165
  methanogenic, 165
Bathtub ring, 37
Bicarbonate, 35–38, 40–42, 50–52
Biological oxygen demand, 165
Blairmore Formation, 2
Blood pressure, 195–198
Blue–green algae, *see* Algae
Boring machine, 4–7, 10
Bridging, 110–112, 144–145
Brine, *see also* Process brine
  draw, 88, 90
  fill, 88–89
  full–strength, 87
  header, 90
  tank, 95

Water Softening with Potassium Chloride: Process, Health, and Environmental Benefits, by William Wist, Jay H. Lehr, and Rod McEachern
Copyright © 2009 by John Wiley & Sons, Inc.

Printed in the United States
By Bookmasters